From Gaia to Selfish Genes

From Gaia to Selfish Genes

Selected Writings in the Life Sciences

edited by Connie Barlow

The MIT Press
Cambridge, Massachusetts
London, England

This book was set in Sabon by The MIT Press and printed and bound in the United States of America.

Library of Congress Cataloging-in-Publication Data

From Gaia to selfish genes : selected writings in the life sciences / edited by Connie Barlow.
p. cm.
Includes bibliographical references (p.)
Includes index.
ISBN 0-262-02323-7
1. Biology—Philosophy. 2. Biology—Research. I. Barlow, Connie C.
QH331.F88 1991
574'.01—dc20 90-13550
CIP

Contents

Preface

The Gaia Hypothesis is for those who like to walk
  or simply stand and stare,
To wonder about the Earth
  and the life it bears
And to speculate about the consequences
                    of our own presence here.

It is an alternative to that pessimistic view which sees
Nature as a primitive force
  to be subdued and conquered.
It is also an alternative to that equally depressing
Picture of our planet as a demented spaceship,
  forever travelling
                    driverless and purposeless
Around an inner circle of the sun.

—James Lovelock

I have taken the liberty of putting to verse one prose paragraph from the 1979 book *Gaia: A New Look at Life on Earth*. James Lovelock is not a poet; he is a scientist. Yet he states his ideas with a clarity and verve that professional writers could emulate. And he is not alone. The other scientists whose works appear in this volume are all skilled and passionate writers as well as respected theorists. Among them, Edward O. Wilson, Douglas Hofstadter, Lewis Thomas, and Richard Dawkins have even received literary awards—including the Pulitzer Prize and the National Book Award.

What a treat to come upon works of literature that address the bold and shocking theories of biology today. Beyond the thousand-and-one facts and the reeking frogs of our school days is a world brimming with ideas, conflicting ideas, that bear on our deepest questions. How truly free is our sense of free will? How autonomous is our sense of individual autonomy? Are we hopelessly naive when we press for a kinder, gentler society? What specifically does biology tell us about the natural bases for competitive or cooperative strategies?

What a piece of work is a man. How noble in reason, how infinite in faculty, in form and moving how express and admirable, in action how like an angel, in apprehension how like a god—the beauty of the world, the paragon of animals!

Hamlet's words may feed our egos. But Shakespeare lived centuries before scientists discovered genes, invented sociobiology, and postulated Gaia. Sigmund Freud in 1917 declared, "The self-love of humanity has been three times severely wounded by the researches of science." First, the Copernican theory cast Earth out of the presumed center of the universe. Next, Charles Darwin showed that "man is not a being different from animals or superior to them." Finally, psychoanalytic theory articulated by Freud himself revealed the power of unconscious mental processes. "The ego is not master in its own house."

Perhaps Freud's list should be expanded today to include a fourth and fifth blow to our collective ego. If Lovelock is on track with his Gaia hypothesis, there exists an organism far more complex and grander than we. Humans may simply be cells (benign or cancerous?) of a single planetary-scale organism. And if Richard Dawkins' selfish gene theory is valid, we are merely "survival machines—robot vehicles blindly programmed to preserve the selfish molecules known as genes."

This book begins with the Gaia hypothesis and ends with the selfish gene theory, making a grand tour of biology from the biggest to the smallest scale. As you will see, each theory has subtle implications, and each has its detractors. Whether this book summons feelings of existential pathos, triggers enlightenment and cosmic peace, or fails to make any impression whatsoever is really up to us. The mindset we embrace and the lessons we derive are ultimately our own choice. Biology does not replace philosophy, but it surely enriches the discourse.

This book is designed to take you on a quest I began five years ago when I chanced upon a lecture by James Lovelock. My childhood fascination for science as nature (which had been utterly quenched by schooling in science as facts) was reignited by Lovelock in the form of science as ideas. I had not an inkling that biology could be this big, this open, this profound. I had no notion that science was so human an endeavor.

For several years thereafter, many of my evenings were devoted to reading popular works written by biologists while energy economics remained my career. One day it occurred to me that other nonscientists (and young, potential scientists) might also enjoy sampling this string of ideas and that I could make the path easier by bringing together the most powerful passages. As my purpose here is not to stump for one worldview or another, you will find a mix of theories and opposing statements in this anthology. I aim to provoke, to encourage casual readers of nonfiction, devotees of popular physics, science majors, and students of the liberal arts to dig into the raft of appealing works in biology. For while physics may have a claim on cosmology and the arts on our imagination, biology speaks to our vision of self, our sense of place on a planet teeming with life.

Note to the Reader

All excerpts from the writings of scientists and science writers will appear in roman type, with authors and sources credited at the beginning of each. *Words that appear in italics are my own editorial insertions or reflect standard usage for italic type.* Chapter titles, headings, and choice of visuals are also generally my own doing, with the concurrence of the authors themselves. I deeply appreciate the flexibility of all authors whose works are excerpted, as their writings often had to be greatly compressed and rearranged. To insure a seamless presentation, deletions are not marked with ellipses. I am also grateful to authors for allowing their material to appear in the same book with that of their critics.

I *Is Earth Itself Alive?*

Every great advance in science has issued from a new audacity of the imagination.

—John Dewey, 1929

The atmosphere is not merely a biological product, but more probably a biological construction: not living, but like a cat's fur, a bird's feathers, or the paper of a wasp's nest, an extension of a living system.

—James Lovelock, 1979

1 *Mother Earth: Myth or Science?*

These excerpts are drawn from two books by James Lovelock, the scientist who envisioned the Gaia hypothesis. Gaia: A New Look at Life on Earth *(copyright 1979 by J. E. Lovelock) is reprinted by permission of the author and Oxford University Press.* The Ages of Gaia *(copyright 1988 by The Commonwealth Fund Book Program of Memorial Sloan-Kettering Cancer Center) is reprinted with permission of the author and W. W. Norton & Company, Inc.*

James Lovelock

Of all the prizes that come from surviving more than fifty years, the best is the freedom to be eccentric. What a joy to be able to explore the physical and mental bounds of existence in safety and comfort, without bothering whether I look or sound foolish. The young usually find the constraints of convention too heavy to escape, except as part of a cult. The middle-aged have no time to spare from the conservative business of living. Only the old can happily make fools of themselves.

The idea that the Earth is alive is at the outer bounds of scientific credibility. I started to think and then to write about it in my early fifties. I was just old enough to be radical without the taint of senile delinquency. My contemporary and fellow villager, the novelist William Golding, suggested that anything alive deserves a name—what better for a living planet than Gaia *(pronounced Gii'-yah)*, the name the Greeks used for the Earth Goddess?

I have frequently used the word Gaia as a shorthand for the hypothesis itself, namely that the biosphere is a self-regulating entity with the capacity to keep our planet healthy by controlling the chemical and physical environment. Occasionally it has been difficult to avoid talking of Gaia as if she were known to be sentient. This is meant no more seriously than is the appellation "she" when given to a ship by those who sail in her, as a recognition that even pieces of wood and metal may achieve a composite identity distinct from the mere sum of its parts.

The concept of Mother Earth has been widely held throughout history and has been the basis of a belief which still coexists with the great religions. As a result of the accumulation of evidence about the natural environment and the growth of the science of ecology, there have recently been speculations that the biosphere may be more than just the complete range of all living things within their natural habitat of soil, sea, and air. Ancient belief and modern knowledge have fused emotionally in the awe with which astronauts with their own eyes and we by indirect vision have seen the Earth revealed in all its shining beauty against the deep darkness of space. Yet this feeling, however strong, does not prove that Mother Earth lives. Like a religious belief, it is scientifically untestable.

Journeys into space did more than present the Earth in a new perspective. They also sent back information about its atmosphere and its surface which provided a new insight into the interactions between the living and the nonliving parts of the planet. From this has arisen the hypothesis in which the Earth's living matter, air, oceans, and land surface form a complex system which can be seen as a single organism and which has the capacity to keep our planet a fit place for life.

Interdisciplinary Pilgrim

This book *(Oxford, 1979)* is a personal account of a journey through space and time in search of evidence with which to substantiate this model of the Earth. The quest began in the mid-1960s and has ranged through territories of many different scientific disciplines, indeed from astronomy to zoology.

Such journeys are lively, for the boundaries between the sciences are jealously guarded by their Professors and within each territory there is a different arcane language to be learnt. In the ordinary way a grand tour of this kind would be extravagantly expensive and unproductive in its yield of new knowledge; but just as trade often still goes on between nations at war, it is also possible for a chemist to travel through such distant disciplines as meteorology or physiology, if he has something to barter.

Usually this is a piece of hardware or a technique. I was fortunate to work briefly with Archer Martin, who developed the technique of gas chromatography. During that time I added some embellishments which extended the range of his invention. One of these was the so-called electron capture detector. This device has exquisite sensitivity in the detection of traces of certain chemical substances. This sensitivity made possible the discovery that pesticides were present in all creatures of the Earth, from penguins in Antarctica to the milk of nursing mothers in the United States. It was this discovery that facilitated the writing of Rachel Carson's influential book, *Silent Spring*, by providing evidence needed to justify her concern over the damage done to the biosphere by the ubiquitous presence of these toxic chemicals.

The electron capture detector has continued to reveal minute but significant quantities of other toxic chemicals in places where they ought not to be. Among these intruders are: PAN (peroxyacetyl nitrate), a toxic component of Los Angeles smog; the PCBs (polychlorobiphenyls) in the remote natural environment; and most recently, in the atmosphere at large, the chlorofluorocarbons and nitrous oxide—substances which are thought to deplete ozone in the stratosphere.

Electron capture detectors enabled me to pursue my quest for Gaia through various scientific disciplines, and also to travel around the Earth itself. Although my role as a tradesman made interdisciplinary journeys feasible, they have not been easy, since these past several decades have witnessed a great deal of turmoil in the life sciences, particularly in areas where science has been drawn into the processes of power politics.

The Search for Life on Mars

The concept that the Earth is actively maintained and regulated by life had its origins in the search for life on Mars. It all started one morning in the spring of 1961 when the postman brought a letter that was for me almost as full of promise and excitement as a first love letter. It was an invitation from NASA to be an experimenter on its first lunar instrument mission.

Space is only a hundred miles away and is now a common place. But 1961 was only four years after the Soviet Union had launched the first artificial satellite, Sputnik. To receive an official invitation to join in the first exploration of the Moon was a legitimization and recognition of my private world of fantasy. My childhood reading had moved on that well-known path from *Grimm's Fairy Tales* through *Alice's Adventures in Wonderland* to Jules Verne and H. G. Wells. I had often said in jest that it was the task of scientists to reduce science fiction to practice. Someone had listened and called my bluff.

My first encounter with the space science of NASA was to visit that open-plan cathedral of science and engineering, the Jet Propulsion Laboratory, just outside the suburb of Pasadena in California. Soon after I began work with NASA on the lunar probe, I was moved to the even more exciting job of designing sensitive instruments that would analyze the surfaces and atmospheres of the planets. My background, though, was biology and medicine, and I grew curious about the experiments to detect life on other planets. I expected to find biologists engaged in designing experiments and instruments as wonderful and exquisitely constructed as the spacecraft themselves. The reality was a disappointment that marked the end of my euphoria. I felt their experiments had little chance of finding life on Mars, even if the planet were swarming with it.

When a large organization is faced with a difficult problem the standard procedure is to hire some experts, and this is what NASA did. This approach is fine if you need to design a better rocket engine. But if the goal is to detect life on Mars, there are no such experts. There were no Professors of Life on Mars, so NASA had to settle for experts of life on Earth. These tended to be biologists familiar with the limited range of living things that they work with in their Earth-bound laboratories.

So the planning of experiments was mostly based on the assumption that evidence for life on Mars would be much the same as for life on Earth. Thus one proposed series of experiments involved an automated microbiological laboratory to sample the Martian soil and judge its suitability to support bacteria, fungi, or other micro-organisms. Additional soil experiments were designed to test for chemicals whose presence would indicate life at work: proteins, amino acids, and substances with the capacity that organic matter has to twist a beam of polarized light in a counter-clockwise direction.

After a year or so, I found myself asking some rather down-to-earth questions, such as, "How can we be sure that the Martian way of life, if any, will reveal itself to tests based on Earth's life style?" To say nothing of more difficult questions, such as, "What is life, and how should it be recognized?"

What Is Life?

Back home in the quiet countryside of Wiltshire, England, after my visits to the Jet Propulsion Laboratories, I had time to do more thinking and reading about the character of life and how one might recognize it anywhere and in any guise. I expected to discover somewhere in the scientific literature a comprehensive definition of life as a physical process, on which one could design life-detection experiments, but I was surprised to find how little had been written about the nature of life itself. The present interest in ecology and the application of systems analysis to biology had barely begun, and there was still in those days the dusty academic air of the classroom about the life sciences.

Data galore had been accumulated on every conceivable aspect of living species, from their outermost to their innermost parts, but in the whole vast encyclopedia of facts the crux of the matter, life itself, was almost totally ignored. At best, the literature read like a collection of expert reports, as if a group of scientists from another world had taken a television receiver home with them and had reported on it. The chemist said it was made of wood, glass, and metal. The physicist said it radiated heat and light. The engineer said the supporting wheels were too small and in the wrong place for it to run smoothly on a flat surface. But nobody said what *it* was.

This seeming conspiracy of silence may have been due in part to the division of science into separate disciplines, with each specialist assuming that someone else has done the job. Some biologists may believe that the process of life is adequately described by some mathematical theorem of physics, and some physicists may assume that it is factually described in the recondite writings on molecular biology which one day he will find time to read. But the most probable cause of our closed minds on the subject is that we already have a rapid, highly efficient life-recognition program in our inherited set of instincts.

Our recognition of living things, both animal and vegetable, is instant and automatic, and our fellow-creatures in the animal world appear to have the same facility. This powerful but unconscious recognition no doubt evolved as a survival factor. Anything living may be edible, lethal, friendly, aggressive, or a potential mate—all questions of prime significance for our welfare and continued existence.

Some of my colleagues at the Jet Propulsion Laboratories mistook my growing skepticism for cynical disillusion and quite properly asked, "Well, what would you do instead?" At that time I could only reply vaguely, "I'd look for an entropy reduction, since this must be a general characteristic of all forms of life." Understandably, this reply was taken to be at the best unpractical and at worst plain obfuscation, for few physical concepts can have caused as much confusion and misunderstanding as has that of entropy.

Entropy is almost a synonym for disorder and yet as a measure of the rate of dissipation of a system's thermal energy, it can be precisely expressed in mathematical terms. It has been the bane of generations of students and is direfully associated in many minds with decline and decay, since its expression in the Second Law of Thermodynamics (indicating that all energy will eventually dissipate into heat universally distributed and will no longer be available for the performance of useful work) implies the predestined and inevitable run-down and death of the Universe.

During this century a few physicists have tried to define life. Bernal, Schroedinger, and Wigner all came to the same general conclusion, that life is a member of the class of phenomena which are open or continuous systems able to decrease their internal entropy (thereby increase order) at the expense of substances or free energy taken in from the environment and subsequently rejected in degraded form. A rough paraphrase might be that life is one of those processes which are found whenever there is an abundant flow of energy. It is characterized by a tendency to shape or form itself as it consumes, but to do so it must always excrete low-grade products to the surroundings.

We can now see that this definition would apply equally well to eddies in a flowing stream, to hurricanes, to flames, or even to refrigerators and many other man-made contrivances. A flame assumes a characteristic shape as it burns, and needs an adequate supply of fuel and air to keep going, and we are now only too well aware that the pleasant warmth and dancing flames of an open fire have to be paid for in the excretion of waste heat and pollutant gases. Entropy is reduced locally by the flame formation, but the overall total of entropy is increased during the fuel consumption.

Atmospheric Clues

Yet even if too broad and vague, this classification of life at least points us in the right direction. It suggests, for example, that there is a boundary or interface between the "factory" area and the surrounding environment which receives the waste products. It also suggests that life-like processes require a flux of energy above some minimal value in order to get going and keep going. Assuming that life on any planet would be bound to use the fluid media—oceans, atmosphere, or both—as conveyor belts for raw materials and waste products, it occurred to me that some of the activity associated with entropy reduction within a living system might spill over into the conveyor-belt regions and alter their composition. The atmosphere of a life-bearing planet would thus become recognizably different from that of a dead planet.

Mars has no oceans. If life had established itself there, it would have had to make use of the atmosphere. Mars therefore seemed a suitable planet for a life-detection exercise based on chemical analysis of the atmosphere. Moreover, this could be carried out regardless of the choice of landing site. Most life-detection experiments are effective only within a target area. Even on Earth, local search techniques would be unlikely to yield much positive evidence of life if the landfall occurred on the Antarctic ice sheet or the Sahara desert or in the middle of a salt lake. But with the atmosphere, any sampling site will do.

While I was thinking on these lines, Dian Hitchcock (a philosopher) visited the Jet Propulsion Laboratories. Her task was to evaluate the logic and information-potential of the many suggestions for detecting life on Mars. The notion of life detection by atmospheric analysis appealed to her, and we began developing the idea together. Using our own planet as a model, we examined the extent to which simple knowledge of the chemical composition of the Earth's atmosphere could provide evidence for life.

Our results convinced us that the only feasible explanation of the Earth's highly improbable atmosphere was that it was being manipulated on a day-to-day basis from the surface, and that the manipulator was life itself. The significant decrease in entropy—or as a chemist would put it, the persistent state of disequilibrium among the atmospheric gases—was on its own clear proof of life's activity.

Our findings and conclusions were, of course, very much out of step with conventional geochemical wisdom in the mid-sixties. Most geochemists regarded the atmosphere as an end-product of planetary outgassing *(the release of gases from deep inside the Earth as the young planet cooled and as volcanoes erupt)* and held that subsequent reactions by abiological *(nonliving)* processes had determined its present state. Oxygen, for example, was thought to come solely from the breakdown of water vapor and the escape of hydrogen into space, leaving an excess of oxygen behind. Our contrasting view required an atmosphere which was a dynamic extension of the biosphere itself. It was not easy to find a journal prepared to publish so radical a notion but, after several rejections, Lynn Margulis and I found an editor, Carl Sagan, prepared to publish it in his journal, *Icarus*.

Nevertheless, considered solely as a life-detection experiment, atmospheric analysis was, if anything, too successful. Even then *(with the help of infrared telescopes)*, enough was known about the Martian atmosphere to suggest that it was mostly carbon dioxide and showed no signs of the exotic chemistry characteristic of Earth's atmosphere. The implication that Mars was probably a lifeless planet was unwelcome news to our sponsors in space research.

Although my tentative suggestion to NASA had been rejected, the idea of looking for a reduction or reversal of entropy as a sign of life had implanted itself in my mind. It grew and waxed fruitful until, with the help of many colleagues, it evolved into the Gaia hypothesis. Thinking about life on Mars gave some of us a fresh standpoint from which to consider life on Earth and led us to formulate a new, or perhaps revive a very ancient, concept of the relationship between the Earth and its biosphere.

Introducing Gaia

Even scientists, who are notorious for their indecent curiosity, shy away from defining life. All branches of formal biological science seem to avoid the question. In the *Dictionary of Biology*, three distinguished biologists succinctly define all manner of words. Under the letter L there is leptotene (the first sign of chromosome pairing in meiosis) and limnology (the study of lakes), but nowhere is life mentioned. The Webster and Oxford dictionaries are not much more help.

To go back to the Webster dictionary, it defines life as: "That property of plants and animals (ending at death and distinguishing them from inorganic matter) which makes it possible for them to take in food, get energy from it, grow, etc." If such manifestly inadequate definitions of life are all I have to work with, can I do much better defining the living organism of Gaia? I have found it very difficult, but if I am to tell you about it I must try.

Living things such as trees and horses and even bacteria can easily be perceived and recognized because they are bounded by walls, membranes, skin, or waxy coverings. Using energy directly from the Sun and indirectly from food, living systems incessantly act to maintain their identity, their integrity. Even as they grow and change, grow and reproduce, we do not lose track of them as visible, recognizable entities.

The name of the living planet, Gaia, is not a synonym for the biosphere. The biosphere is defined as that part of the Earth where living things normally exist. Still less is Gaia the same as the biota, which is simply the collection of all individual living organisms. The biota and the biosphere taken together form part but not all of Gaia. Just as the shell is part of a snail, so the rocks, the air, and the oceans are part of Gaia. Gaia has continuity with the past back to the origins of life, and extends into the future as long as life persists. Gaia, as a total planetary being, has properties that are not necessarily discernible by just knowing individual species or populations of organisms living together.

The Gaia Hypothesis

The Gaia hypothesis, when we introduced it in the 1970s, supposed that the atmosphere, the oceans, the climate, and the crust of the Earth are regulated at a state comfortable for life because of the behavior of living organisms. Specifically, the Gaia hypothesis said that the temperature, oxidation state, acidity, and certain aspects of the rocks and waters are at any time kept constant, and that this homeostasis is maintained by active feedback processes operated automatically and unconsciously by the biota. The conditions are only constant in the short term and evolve in synchrony with the changing needs of the biota as it evolves. Life and its environment are so closely coupled that evolution concerns Gaia, not the organisms or the environment taken separately.

You may find it hard to swallow the notion that anything as large and apparently inanimate as the Earth is alive. Surely, you may say, the Earth is almost wholly rock and nearly all incandescent with heat. I am indebted to Jerome Rothstein, a physicist, for his enlightenment on this, and other things. In a thoughtful paper on the living Earth concept (given at a symposium held in the summer of 1985 by the Audubon Society) he observed that the difficulty can be lessened if you let the image of a giant redwood tree enter your mind. The tree undoubtedly is alive, yet 99 percent is dead. The great tree is an ancient spire of dead wood, made of lignin and cellulose by the ancestors of a thin layer of living cells. How like the Earth, and more so when we realize that many

"The atmosphere is not merely a biological product, but more probably a biological construction: not living, but like a cat's fur, a bird's feathers, or the paper of a wasp's nest, an extension of a living system designed to maintain a chosen environment."—James Lovelock

of the atoms of the rocks far down into the magma were once part of the ancestral life from which we all have come.

The first clue to the presence of Gaia was the chemistry of Earth's atmosphere.

The chemical composition of the atmosphere bears no relation to the expectations of steady-state chemical equilibrium. The presence of methane, nitrous oxide, and even nitrogen in our present oxidizing atmosphere represents violation of the rules of chemistry to be measured in tens of orders of magnitude. Disequilibria on this scale suggest that the atmosphere is not merely a biological product, but more probably a biological construction: not living, but like a cat's fur, a bird's feathers, or the paper of a wasp's nest, an extension of a living system designed to maintain a chosen environment.

Mars, Venus, and Earth

When the Earth was first seen from outside and compared as a whole planet with its lifeless partners, Mars and Venus, it was impossible to ignore the sense that the Earth was a strange and beautiful anomaly. The questions raised by space science were at first narrowly focused on a practical question: How is life on another planet to be recognized? Because that question could not be explained solely by conventional biology or geology, I became preoccupied with another question: What if the difference in atmospheric composition between the Earth and its neighbors Mars and Venus is a consequence of the fact that the Earth alone bears life?

The least complex and most accessible part of a planet is its atmosphere. Long before the Viking spacecraft landed on Mars or the Russian Venera landed on Venus, we knew the chemical compositions of their atmospheres. In the middle 1960s, telescopes tuned to the infrared radiation reflected by the molecules of atmospheric gases were used to view Mars and Venus. These observations revealed the identity and proportion of the gases with fair accuracy.

Mars and Venus both had atmospheres dominated by carbon dioxide, with only small proportions of oxygen and nitrogen. More important, both had atmospheres close to the chemical equilibrium state; if you took a volume of air from either of those planets, heated it to incandescence in the presence of a representative sample of rocks from the surface, and then allowed it to cool slowly, there would be little or no change in composition after the experiment.

The Earth, by contrast, has an atmosphere dominated by nitrogen and oxygen. A mere trace of carbon dioxide is present, far below the expectation of planetary chemistry. There are unstable gases such as nitrous oxide, and gases such as methane that react readily with the abundant oxygen. If the same heating-and-cooling experiment were tried with a sample of the air that you are now breathing, it would be changed. It would become like the atmospheres of Mars and Venus: carbon-dioxide dominant, oxygen and nitrogen greatly diminished, and gases such as nitrous oxide and methane absent. It is not too far-fetched to look on the air as like the gas mixture that enters the intake of an internal combustion engine: combustible gases, hydrocarbons, and oxygen mixed. The atmospheres of Mars and Venus are the exhaust gases, all energy spent.

The amazing improbability of the Earth's atmosphere reveals negentropy (negative entropy) and the presence of the invisible hand of life. Take for example oxygen and methane. Both are present in our atmosphere in constant quantities; yet in sunlight they react chemically to give carbon dioxide and water vapor. Anywhere you travel on the Earth's surface to measure it, the methane concentration is about 1.5 parts per million. Close to 1,000 million tons of methane must be introduced into the atmosphere annually to maintain methane at a constant level. In addition, the oxygen used in oxidizing the methane must be replaced—at least 2,000 million tons yearly. The only feasible explanation for the persistence of this unstable atmosphere at a constant composition, and for periods vastly longer than the reaction times of its gases, is the influence of a control system, Gaia.

Who replenishes atmospheric oxygen and methane? Green plants are the source of oxygen, as oxygen is a waste product of photosynthesis. Methane is a waste product of certain kinds of bacteria that reside in oxygen-poor (anaerobic) habitats—notably in the muds of freshwater and saltwater marshes and in the guts of practically all animals that feed on fibrous plants. Without a healthy population of these friendly microbes, wood-eating termites, grass-eating cattle, and bean-eating humans would be unable to digest their food. To put this concept rather indelicately, every time you fart, you provide a little more atmospheric evidence that there is indeed life on Earth.

A lifeless Earth would have an atmosphere just like that of Mars and Venus. Oxygen would be a mere trace of what it is now on Earth; nitrogen would be gone largely into the seas; and methane, hydrogen, and ammonia would vanish in just a few years. When the air, the ocean, and the crust of our planet are examined in this way, the Earth is seen to be a strange and beautiful anomaly.

Climate Control

Lovelock recognized a second clue to the presence of Gaia when he compared what paleontologists know about the history of life (recorded in fossils) with the conclusions drawn by astrophysicists about the history of our sun.

The history of the Earth's climate is one of the more compelling arguments in favor of Gaia's existence. We know from the record of the sedimentary rocks that for the past 3.5 billion years the climate has never been, even for a short period, wholly unfavorable for life. Because of the unbroken record of life, we also know that the oceans can never have either frozen or boiled. Indeed, evidence in the rocks strongly suggests that the climate has always been much as it is now, except during glacial periods or near the beginning of life when it was somewhat warmer. The glacial cold spells—Ice Ages, as they are called, often with exaggeration—affected only those parts of the Earth outside latitudes 45° North and 45° South. We are inclined to overlook the fact that 70 percent of the Earth's surface lies between these latitudes. The so-called Ice Ages only affected the plant and animal life which had colonized the remaining 30 percent, which is often partially frozen even between glacial periods, as it is now. Extinction through glaciation was not the only danger. Overproduction of ammonia, carbon-dioxide, and other heat-retaining gases could have resulted in the runaway greenhouse effect—that is, to a rapidly increasing surface temperature that would have scorched the Earth and left it permanently lifeless, as is the planet Venus now.

Before there was a significant amount of oxygen in the air, the emission and absorption of ammonia by simple organisms may have been the temperature control process. Later, when photosynthesizing and respiring organisms existed and oxygen became a major part of the air, the control of carbon dioxide, which is also a heat absorbing gas, may have played a role in stabilizing temperature.

We may at first think that there is nothing odd about this picture of a stable climate over the past 3.5 billion years. The Earth had no doubt long since settled down in orbit around that great radiator, the sun, so why should we expect anything different? Yet it is odd, and for this reason. Our sun, being a typical star, has evolved according to a standard and well established pattern. A consequence of this is that during the 3.5 billion years of life's existence on the Earth, the sun's output of energy increased by at least 30 percent. If the Earth's climate were determined solely by the output from the sun, our planet would have been frozen during the first 1.5 billion years of life's existence. We know from the record of the rocks and from the persistence of life itself that no such adverse conditions existed.

The lack of warmth of a cooler Sun could have been offset by a blanket of greenhouse gas. Gases with more than two atoms in their molecules have the interesting property of absorbing the radiant warmth, the infrared radiation, that escapes from the Earth's surface. These gases, which include carbon dioxide, water vapor, and ammonia, are transparent to the visible and the almost visible infrared radiation. These are the parts of the Sun's spectrum that carry the most energy; radiant heat in this form will penetrate the air and warm the surface. The same gases are opaque to the longer wavelength infrared that radiates from the Earth's surface and lower atmosphere. The trapping of the warmth, which otherwise would escape to space, is the greenhouse effect; so called because it is like, although not the same as the warming effect of the glass panes of a greenhouse.

The climate and the chemical properties of the Earth now and throughout its history seem always to have been optimal for life. For this to have happened by chance is as unlikely as to survive unscathed a drive blindfold through rush-hour traffic.

Control without Consciousness

To many scientists, Gaia was a teleological concept, one that required foresight and planning by the biota. How in the world could the bacteria, the trees, and the animals have a conference to decide optimum conditions? How could organisms keep oxygen at 21 percent and the mean temperature at 20°C? Not seeing a mechanism for planetary control, they denied its existence as a phenomenon and branded the Gaia hypothesis as teleological. This was a final condemnation. Teleological explanations are a sin against the holy spirit of scientific rationality; they deny the objectivity of Nature.

Thus, the most persuasive criticism of Gaia theory is that planetary homeostasis, by and for living organisms, is impossible because it would require the evolution of communication between the species and a capacity for foresight and planning. The critics who made this challenging, and to me helpful, criticism were not concerned with the practical evidence that the Earth has kept a climate favorable for life in spite of major perturbations, or that the atmosphere is now stable in its composition in spite of the chemical incompatibility of its component gases. They were criticizing from the certainty of their knowledge of biology. No organism as large, and, as they saw her, sentient, could possibly exist. I think this criticism is dogmatic, and it is easy to answer.

It is a tribute to the success of biogeochemistry that most Earth scientists today agree that the reactive gases of the atmosphere are biological products. But most would disagree that the biota in any way *control* the composition of the atmosphere, or any of the important variables, such as global temperature and oxygen concentration. I knew that there was little point in gathering more evidence about the capacity of the Earth to regulate its climate and composition. Mere evidence by itself could not be expected to convince mainstream scientists that the Earth was regulated by life. Scientists usually want to know how it works; they want a mechanism. What was needed was a Gaian model.

A Computer Model for Gaia

In what way do Gaian models differ from the conventional biogeochemical ones? Does the assumption of the close coupling of life and its environment change the nature of the whole system? Is homeostasis a reasonable prediction of Gaia theory?

The difficulty in answering these questions comes from the sheer complexity of the biota and the environment, and because they are interconnected in multiple ways. There is hardly a single aspect of their interaction that we can confidently describe by a mathematical equation. A drastic simplification was needed. I wrestled with the problem of reducing the complexity of life and its environment to a simple scheme that could enlighten without distorting. Daisyworld was the answer. I first described this model in 1982 at a conference in Amsterdam, and published a paper, "The Parable of Daisyworld," in *Tellus* in 1983 with my colleague Andrew Watson. I am indebted to Andrew for the clear, graphic way of expressing it in formal mathematical terms in this paper.

Picture a planet about the same size as the Earth, spinning on its axis and orbiting, at the same distances as the Earth, a star of the same mass and luminosity as the Sun. This planet differs from the Earth in having more land area and less ocean, but it is well watered, and plants will grow almost anywhere on the land surfaces when the climate is right. This is the planet Daisyworld, so called because the principal plant species are daisies of different shades of color: some dark, some light, and some neutral colors in between.

The star that warms and illuminates Daisyworld shares with our Sun the property of increasing its output of heat as it ages. When life started on Earth about 3.8 billion years ago, the Sun was about 30 percent less luminous than now. In a few more billion years, it will become so fiercely hot that all life that we know will die or be obliged to find another home planet. The increase of the Sun's brightness as it ages is a general and undoubted property of stars. As the star burns hydrogen (its nuclear fuel) helium accumulates. The helium, in the form of a gaseous ash, is more opaque to radiant energy than is hydrogen and so impedes the flow of heat from the nuclear furnace at the center of the star. The central temperature then rises and this in turn increases the rate of hydrogen burning until there is a new balance between heat produced at the center and the heat lost from the solar surface. Unlike ordinary fires, star-sized nuclear fires burn fiercer as the ash accumulates and sometimes even explode.

Daisyworld is simplified in the following ways. The environment is reduced to a single variable, temperature, and the biota to a single species, daisies. If too cold, below 5°C, daisies will not grow; they do best at a temperature near 20°C. If the temperature exceeds 40°C, it will be too hot for the daisies, and they will wilt and die. The mean temperature of the planet is a simple balance between the heat received from the star and the heat lost to the cold depths of space in the form of long-wave infrared radiation. On the Earth, this heat balance is complicated by the effects of clouds and of gases such as carbon dioxide. The sunlight may be reflected back to space by the clouds before it can reach and warm the surface. On the other hand, the heat loss from the warm surface may be lessened because clouds and molecules of carbon dioxide reflect it back to the surface. Daisyworld is assumed to have a constant amount of carbon dioxide, enough for daisies to grow but not so much as to complicate the climate. Similarly, there are no clouds in the daytime to mar the simplicity of the model, and all rain falls during the night.

The mean temperature of Daisyworld is, therefore, simply determined by the average shade of color of the planet, or as astronomers call it, the albedo. If the planet is a dark shade, low albedo, it absorbs more heat from the sunlight and the surface is warmed. If light in color, like fallen snow, then 70 or 80 percent of the sunlight may be reflected back to space. Such a surface is cold when compared with a dark surface under comparable solar illumination.

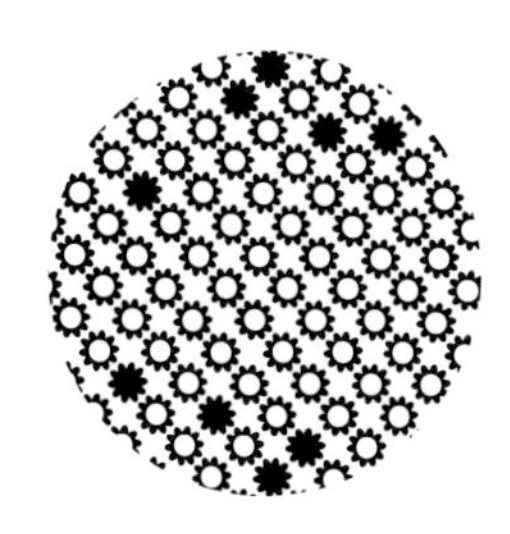

Exploring Daisyworld

Imagine a time in the distant past of Daisyworld. The star that warms it was less luminous, so that only in the equatorial region was the mean temperature of bare ground warm enough, 5°C, for growth. Here daisy seeds would slowly germinate and flower. Let us assume that in the first crop multicolored, light, and dark species were equally represented. Even before the first season's growth was over, the dark daisies would have been favored. Their greater absorption of sunlight in the localities where they grew would have warmed them above 5°C. The light-colored daisies would be at a disadvantage. Their white flowers would have faded and died because, reflecting the sunlight as they do, they would have cooled below the critical temperature of 5°C.

The next season would see the dark daisies off to a head start, for their seeds would be the most abundant. Soon their presence would warm not just the plants themselves, but, as they grew and spread across the bare ground, would increase the temperature of the soil and air, at first locally and then regionally. With this rise of temperature, the rate of growth, the length of the warm season, and the spread of dark daisies would exert a positive feedback and lead to the colonization of most of the planet by dark daisies. The spread of dark daisies would eventually be limited by a rise of global temperature to levels above the optimum for growth. Any further spread of dark daisies would lead to a decline in seed production. In addition, when the global temperature is high, white daisies will grow and spread in competition with the dark ones. The growth and spread of white daisies is favored then because of their natural ability to keep cool.

As the star that shines on Daisyworld grows hotter and hotter, the proportion of dark to light daisies changes until, finally, the heat flux is so great that even the whitest daisy crop cannot keep enough of the planet below the critical 40°C upper limit for growth. At this time flower power is not enough. The planet becomes barren again, and so hot that there is no way for daisy life to start again.

It is easy to make a numerical model of Daisyworld simple enough to run on a personal computer. Daisy populations are modeled by differential equations borrowed from theoretical ecology. When I first tried the Daisyworld model I was surprised and delighted by the strong regulation of planetary temperature that came from the simple competitive growth of plants with dark and light shades. I did not invent these models because I thought that daisies, or any other dark- and light-colored plants, regulated the Earth's temperature by changing the balance between the heat received from the Sun and that lost to space. I had designed them to answer the criticism of Ford Doolittle and Richard Dawkins that Gaia was teleological. In Daisyworld, one property of the global environment, temperature, was shown to be regulated effectively, over a wide range of solar luminosity, by an imaginary planetary biota without invoking foresight or planning. This is a definitive rebuttal of the accusation that the Gaia hypothesis is teleological, and so far it remains unchallenged.

The simple model Daisyworld illustrated how Gaia might work. No foresight, planning, or purpose was invoked. Daisyworld is a theoretical view of a planet in homeostasis. We can now begin to think of Gaia as a theory, something rather more than the mere "let's suppose" of an hypothesis.

Gaia and Evolution

So what is Gaia? If the real world we inhabit is self-regulating in the manner of Daisyworld, and if the climate and environment we enjoy and freely exploit is a consequence of an automatic, but not purposeful, goal-seeking system, then Gaia is the largest manifestation of life. The outer boundary is the Earth's atmospheric edge to space. The boundary of the planet then circumscribes a living organism, Gaia, a system made up of all the living things and their environment.

There is no clear distinction anywhere on the Earth's surface between living and nonliving matter. There is merely a hierarchy of intensity going from the "material" environment of the rocks and the atmosphere to the living cells. But at great depths below the surface, the effects of life's presence fade. It may be that the core of our planet is unchanged as a result of life; but it would be unwise to assume it.

Gaia as the largest manifestation of life differs from other living organisms of Earth in the way that you or I differ from our population of living cells. At some time early in the Earth's history before life existed, the solid Earth, the atmosphere, and oceans were still evolving by the laws of physics and chemistry alone. It was careering, downhill, to the lifeless steady state of a planet almost at equilibrium. Briefly, in its headlong flight through the ranges of chemical and physical states, it entered a stage favorable for life. At some special time in that stage, the newly formed living cells grew until their presence so affected the Earth's environment as to halt the headlong dive towards equilibrium. At that instant the living things, the rocks, the air, and the oceans merged to form the new entity Gaia. Just as when the sperm merges with the egg, new life was conceived.

Gaia is no static picture. She is forever changing as life and the Earth evolve together, but in our brief life span she keeps still long enough for us to begin to understand and see how fair she is. The evolution of *Homo sapiens*, with his technological inventiveness and his increasingly subtle communications network, has vastly increased Gaia's range of perception. She is now through us awake and aware of herself. She has seen the reflection of her fair face through the eyes of astronauts and the television cameras of orbiting spacecraft. Our sensations of wonder and pleasure, our capacity for conscious thought and speculation, our restless curiosity and drive are hers to share.

This new interrelationship of Gaia with man is by no means fully established; we are not yet a truly collective species, corralled and tamed as an integral part of the biosphere, as we are as individual creatures. It may be that the destiny of mankind is to become tamed, so that the fierce, destructive, and greedy forces of tribalism and nationalism are fused into a compulsive urge to belong to the commonwealth of all creatures which constitutes Gaia. It might seem to be a surrender, but I suspect that the rewards, in the form of an increased sense of well-being and fulfillment, in knowing ourselves to be a dynamic part of a far greater entity, would be worth the loss of tribal freedom.

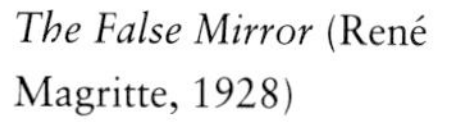

The False Mirror (René Magritte, 1928)

Peacocks and Spectroscopes: Profile of James Lovelock

Who is James Lovelock? In November 1986 the New York Times Magazine *published a profile of Lovelock, written by Lawrence E. Joseph. Joseph is a science writer whose 1990 book,* Gaia: The Growth of an Idea, *is a fascinating account of the people and events driving the science of Gaia. Below are a few excerpts from the earlier magazine piece (reprinted here by permission of the author) revealing Lovelock's unusual lifestyle, which blends high-technology science with the peaceful tone of a country home—Daisyworld indeed!*

Lawrence E. Joseph

Discovering a nest of gleaming technology deep in the countryside can be a delightful surprise, much like finding a flower bed in the middle of a busy city. Not far from a river in St. Giles-on-the-Heath, a hamlet on the Cornwall-Devon border in southwestern England, stands the laboratory of James E. Lovelock, a white, windowed cabin attached to his house and patrolled outside by half a dozen peacocks. A phalanx of spectroscopes, radiation detectors and microcomputers set incongruously in a fragrant meadow, the lab seems like a probe sent to unravel the secrets of nature.

A maverick honored by the establishment, Lovelock is one of the very few independent, unaffiliated scientists to become a Fellow of the Royal Society, the British counterpart of America's National Academy of Sciences. Generally considered a biologist, Lovelock is an interdisciplinary wanderer through the jealously specialized halls of academe; he holds a Ph.D. in medicine from the University of London and was professor of chemistry at the Baylor College of Medicine in Houston and a visiting professor of cybernetics—the study of controlled systems—at the University of Reading in England. Recently, he was elected president of the Marine Biological Association of the United Kingdom.

Above all, Lovelock considers himself an inventor; royalties on his many patents and other scientific devices, and retainers from Hewlett-Packard, Shell and other major corporations, for decades have supported his research as well as his wife and four children. His most important invention is the electron capture detector. First developed in 1957, it is still the most sensitive instrument around for analyzing the air.

Lovelock has even been the co-author of a novel, *The Greening of Mars*, with Michael Allaby, published in 1984. The book is science "faction," a speculative work supported by scientific principles; in it, humans create an earthlike environment on the red planet by infusing its atmosphere with chlorofluorocarbon gases, the same ones at the center of the current controversy about the ozone layer. As a result, they are able to create a new civilization. "Did you know that *The Greening of Mars* spun off no less than three major scientific meetings within a year of its publication?" asks Lovelock. "One of them held by the Royal Society of Canada, no less."

The scientist leaves the lab and crosses the bridge he has built over the river that divides his property. "One cannot overestimate the power of a good story," says Jim Lovelock, as he wanders into a meadow profuse with wild flowers and lush grass.

The Scientist as Artist

Lovelock's life in the Cornwall countryside may sound idyllic to readers who share his passion for nature and independence, but getting there was not so easy. In a revealing article titled "The Independent Practice of Science" (which appeared in the 6 September 1979 issue of New Scientist *magazine, London) Lovelock recalls the rigors.*

James Lovelock

The artist, the novelist and the composer of music all tend to lead solitary lives, often working from their own homes. Scientists almost always work in university departments, government or private institutions. A solitary one is not only unusual. He may be suspect as mad or rejected as irresponsible. The independent practice of science does not happen by accident, and the events in my life which led me towards this unusual style of working need some explanation.

I left school in London early in 1938. The parting words of my headmaster were in those days wise but none-the-less unheeded. He said: "Lovelock, you are a fool to take up science. There is no place there except for those of genius or with private means." He well knew that I had neither of these nor any prospect of acquiring them. His words were well meant and at the time very true, for in those post-depression years, just before the Second World War, newly graduated chemists were employed for a year or more in some notorious industries at no salary on the excuse that they were acquiring industrial experience. But I was not only obstinate in my determination to do science of any sort. I was also incredibly lucky.

My first employer, Humphrey Desmond Murray, was a kindly and tolerant man and his business as consultant chemist covered topics from the advanced organic chemistry of the synthesis of new developing agents in color photography to the invention of invisible but detectable powders with which to mark bank notes for Scotland Yard. He provided what in those days was a princely wage—£3 per week. He also paid my fees to attend that splendid institution, Birkbeck College, London, at which to attain a degree by evening class study.

Most artists would think that there was nothing unusual in this way of life. It was that of apprenticeship, learning the profession at the hands of a master. But strangely, somehow, then and now apprenticeship is regarded as inferior to full time university training. And this is where I think that science at sometime in the past took a wrong turning.

In 1939, with the outbreak of the Second World War, Birkbeck College was closed but I was able to complete my undergraduate training as a full time student at Manchester. University seemed very unreal after the rich experience of working life and I could hardly wait for the time to come when I could do rather than be taught how. And this is in no way meant to discredit the lecturers at Manchester for they were very good indeed. It was rather that there are some of us better able to teach ourselves heuristically, than to be taught. For us the years from puberty to the completion of university training are like a prison sentence. I feel for those now who have so many more years to serve than I did then.

After receiving his degree, Lovelock spent two decades holding down more or less normal jobs with various research institutions—all the while plotting and dreaming for a life of independent research. Finally his various successes gave him a reputation sufficient to proceed on his own. He retreated to a thatched cottage in the countryside with his wife and children, and there began the independent practice of science. Difficult times were still ahead, however.

One of the most frustrating experiences was when a grant application was turned down because the reviewer concluded, "Every schoolboy knows that the chlorofluorocarbons are among the most stable of chemical substances. It would be very difficult to measure them in the atmosphere at parts per million on account of their lack of chemical reactivity. The proposer suggests that he can measure them at parts per billion; the application is clearly bogus and the time of this committee should not be wasted with such frivolous proposals." Lovelock was not deterred. Despite lack of funds, he schemed his way onto a transoceanic research vessel and there made the measurements—in the parts per billion—that provided other scientists with the essential data they needed to discern the grave problems facing our atmospheric ozone shield.

What drove Lovelock to pursue this difficult dream of independence?

My ultimate objective was to work like the artist or the novelist. I wanted to be able to do scientific creative work without any constraints from employers or customers who, even with the best will in the world, tend to interfere. Gradually I found that the answer lay in doing only those things which were truly interesting or which curiosity inspired. Strangely, the work done to answer such questions as "I wonder if" almost always led to bread and butter in two or three years time. Not so the dull and pedestrian questions which tended to be as limited in their prospects as their scope. A two or three year wait is not long and no worse than the time span a novelist must contend with between the first idea and the first receipt of royalties from a book he has written and has published.

Lovelock, however, was practical too. Although his own Gaia hypothesis has been "a source of inspiration" for more than two decades now, he did work on other things.

The artist all too often finds no market for the work he likes to do and is consequently obliged to produce "potboilers" to survive. So it is with independent scientists—except that our pot-boilers are inventions. I have made about thirty of these, which from time to time were patented and sold and the proceeds used to sustain the family and the laboratory.

For inventors and scientists themselves, so long as there is enough to eat, warmth to work in, and a family prepared to share, relative poverty is more of a spur than a deterrent. It is probably more difficult to work as an independent in the United States, where the abyss of ruin through the penal cost of ill health forever causes anxiety. For me and my family the more compassionate social scene of the United Kingdom has made possible our independent life. Otherwise chronic ill-health and disability, which though no physical deterrent in itself, would have been elsewhere an impossible financial burden.

The most unexpected and in some ways most satisfying aspect of independence is the very real sense of security it provides. A man or a woman with a permanent job and indexed pension rights would seem to have no worries and be free to concentrate on the work in hand. But perversely this is not so. It is the total uncertainty about the next year's job and next year's income that makes every day a possible new adventure. Indeed, it makes next year a lifetime ahead and no cause for concern today.

2 *The Search for Gaia*

In his first book, Gaia: A New Look at Life on Earth *(1979), James Lovelock explored the ways in which life could have been involved in regulating both the chemistry and the climate of the Earth. But at the time of publication Lovelock could not identify a test for determining whether any of these plausible mechanisms were indeed real and sustaining a livable world today. And while his later model of Daisyworld expressed a powerful logic as to how a Gaian system could regulate itself without really knowing what it was doing, without being conscious, Lovelock had offered no concrete examples of the mechanisms of such regulation. Who among us is playing the role of the black and white daisies in regulating the albedo of our very real planet? And what organisms hold the portfolio for maintaining atmospheric chemistry, ocean salinity, and so on?*

While Lovelock's ideas warmed the hearts of many who look to Nature for spiritual meaning and connection, the Gaia hypothesis appeared doomed to the fringes of science. Then, in the late 1980s, Lovelock was part of a team of scientists that discovered the first possible Gaian mechanism. Excerpts here are drawn from an article by science writer Richard Monastersky. "The Plankton-Climate Connection" appeared in the 5 December 1987 issue of Science News *and is reprinted with permission from Science Service, Inc. (copyright 1987).*

Richard Monastersky

For more than 3.5 billion years, the earth has provided the exact mix of water, gases and temperature that life, as we know it, requires for existence. Is this a fortuitous occurrence? Has life been riding around on a huge piece of rock that has luckily furnished suitable accommodation? Or has life taken part in the process, creating for itself a cozy environment that can support plankton, pandas and a panoply of other earthly organisms?

These are the questions raised by a controversial theory called the Gaia hypothesis. Proposed in the early 1970s by British scientist James E. Lovelock, working with biologist Lynn Margulis, the Gaia hypothesis states that life has regulated and stabilized the environment, keeping it within the narrow bounds that allow life to continue. And now, microscopic marine plants are floating into the Gaia debate.

Plankton, Clouds, and Climate

Evidence is emerging that these one-celled marine plants, called plankton, are at least partially responsible for the temperature of the earth. And the new findings are enhancing scientists' understanding of how life can influence the global climate. Scientists are just beginning to understand that plankton and climate are linked, and the exact terms of the relationship are still unclear. But what excites many researchers is the possibility that this might be an example of a Gaian relationship. Perhaps, say Gaia supporters, these plants behave like a living thermostat, maintaining the earth's temperature at a comfortable level—a suggestion that is stimulating both experiment and debate.

How is it that such tiny creatures can affect the climate? It turns out that certain species of plankton produce a chemical compound that is important in the formation of clouds. And the number of clouds affects the earth's climate because clouds reflect much of the sun's radiation, sending it back into space. Airplane travelers who have looked out on a blindingly bright field of white are well acquainted with this reflectivity.

The chemical link between plankton and clouds is a sulfur compound called dimethylsulfide (DMS). Inside plankton cells, DMS protects against the high salt concentrations of the outside seawater. The cells excrete DMS as a normal metabolic product, but the compound also enters seawater when plankton are eaten or die. DMS then diffuses from the seawater into the atmosphere.

For decades, says James Lovelock, this was the extent of the scientific knowledge about DMS. Nobody really questioned the subsequent fate of the compound. "They thought it was just a biochemical curiosity," he says. But in the 1970s, research into the Gaia hypothesis prompted Lovelock to examine the world's sulfur budget in detail, which meant keeping close track of all the important forms of sulfur. And this continued interest in DMS helped reveal its climatic role.

While Lovelock was musing on the fate of DMS, Robert Charlson of the University of Washington in Seattle was working independently at the other end of the problem. As an atmospheric scientist, Charlson had been wondering how marine clouds grow. Clouds are actually composed of particles of water. But before they can develop, there must be some sort of nuclei onto which water can condense. Says Lovelock, "It was speculation by Bob Charlson: Where did the cloud-condensation nuclei come from? We talked about DMS, and it seemed natural that it could be the source."

In the April 16, 1987 issue of *Nature*, Lovelock, Charlson and their colleagues proposed that once DMS reaches the atmosphere, it oxidizes to form sulfate particles, which then serve as condensation nuclei for developing cloud particles. According to the theory, if plankton produced more DMS, the surface of the earth would get cooler. New evidence has confirmed the major links in this plankton-climate chain.

The nineteenth-century naturalist Ernst Haeckel investigated the stunning beauty of nature revealed at the smallest scale. In addition to these "tests" of radiolarians, he drew other microscopic creatures, including diatoms and foramanifera, that comprise the foundation of the marine food chain: plankton. Radiolarians and diatoms both build their structural support and armor from the stuff of sand: silica. Foramanifera and coccolithophorids (the latter are a major producer of DMS) build their shells from calcium carbonate, as do clams and oysters.

Charlson, with Timothy S. Bates and Richard H. Gammon of the Pacific Marine Environmental Laboratory in Seattle, reported in the September 24, 1987 *Nature* that they found a direct correlation between DMS levels and the numbers of cloud nuclei in portions of the Pacific.

The next step in the plankton-climate theory linked the number of condensation nuclei to the reflectivity of clouds. Scientists have understood this theoretical relationship for years. If a cloud has a certain amount of water and the number of available nuclei increases, explains Bates, "it means that the given water vapor is spread out more, and you have more surface area to reflect incoming solar radiation."

But that relationship had gone untested until this year, when James A. Coakley and his colleagues noticed that the exhaust from ships left long, highly reflective trails in marine clouds—an effect that showed up on satellite images. Because exhaust is chiefly made up of different kinds of particles, the observation showed that an increase in particles over the ocean could raise the reflectivity of clouds, says Coakley, of the National Center for Atmospheric Research (NCAR) in Boulder, Colorado. He reported his findings in the August 28, 1987 issue of *Science*.

The Search for Feedback Loops

The recent findings have prompted Lovelock, Charlson, and their colleagues to wonder whether the plankton-climate connection is an example of a Gaia-like process. To answer that question, they need to determine whether the relationship involves what scientists call a negative feedback loop.

Feedback loops are "circular" processes in which two elements of a system affect each other simultaneously. They come in two varieties: negative and positive. Negative feedback reduces any changes in a system and therefore helps to stabilize it. If negative feedback were active in the plankton-climate relationship, then a shift toward warmer temperatures would stimulate plankton to produce more DMS. This would create more clouds, bringing temperatures back down. Positive feedback, on the other hand, helps destabilize a system. In this process, warmer temperatures would cause plankton to reduce DMS production, which would warm the earth even more. While negative feedback works as a thermostat, positive feedback works like an amplifier.

Scientists from varied disciplines are now attempting to quantify the two-way relationship between plankton and the climate. Through ocean observations, ice core data from Antarctica and laboratory experiments, researchers hope to determine when this system may have served to regulate temperatures, and perhaps when it has destabilized temperatures, spurring the spread of ice over large sections of the earth.

For those involved in the Gaia debate, the plankton-climate relationship represents a chance to study feedback on a refreshingly concrete system. In the past two decades, many scientists have complained that

Sky Above Clouds IV (Georgia O'Keeffe, 1965)

the examples raised in support of Gaia are untestable and far too general. Says Gaia coauthor Lynn Margulis, "This kind of work does not solve or prove the Gaia hypothesis, but it provides mechanisms which can be tested."

Concerning the proposed plankton-climate relationship, Stephen Schneider, an NCAR climatologist and sometimes Gaia critic, says, "I like it a lot. It's testable, which is certainly not true of a lot of other parts of Gaia. And it's clear cut: We can measure it. It may not be fundamental; we don't know yet. But it's a nice link and I think it should be pursued."

Truth or Utility?

Whether scientists accept or reject the Gaia theory is not a major concern of Lovelock's. He says he's primarily delighted that it is stimulating research. "All that matters at this stage of the game," he says, "is that they don't just ignore it, and that they go out and experiment and try to prove it wrong." As a case in point he mentions that it was the Gaia theory that prompted scientists to discover the relationship between plankton and climate. "What's exciting about all this is that by following our noses along a certain trail, we're finding a lot of interesting science."

In the spring of 1988 I attended a conference on the Gaia hypothesis. It was the first such conference sponsored by a reputable scientific society: the American Geophysical Union. A few invited scientists refused to attend; they felt that Lovelock's hypothesis was more fantasy than science. But those who did attend discovered that scientists in fields diverse as geology, oceanography, climatology, microbiology, and ecology had been stimulated by Lovelock's ideas to look for (and find) new connections between the living and nonliving worlds. By the end of the conference most participants had stopped arguing whether the Gaia hypothesis could or could not be tested, whether the notion of a living Earth was science or science fiction. One participant spoke thus:

Science never tells us what to study. Ultimately, that is something that must come from outside of science. To me, the concept of Gaia is most useful as a way of guiding our thinking about what we study, why we study it, and what hypotheses to test. Gaia is not a hypothesis itself; it suggests things that are hypotheses.

Science writer Francesca Lyman captures this perspective in the article "What Gaia Hath Wrought," which appeared in the July 1989 issue of Technology Review. *The excerpts here are reprinted with permission from Technology Review, copyright 1989.*

Gaia and Earth Systems Science

Francesca Lyman

Not until last year, 16 years after the Gaia hypothesis was first proposed, did it receive a hearing from the scientific establishment, as the subject of the distinguished Chapman Conference, sponsored by the American Geophysical Union (AGU). Stephen Schneider, a climatologist at the National Center for Atmospheric Research who thought scientists had not examined Gaia closely enough, organized the event. There was considerable resistance, comments Schneider, who has himself criticized parts of the theory. "I was condemned as 'embarrassing science by putting on that nonsensical meeting,'" he notes. In the end, though, conferees gave Lovelock a standing ovation. Even critics said he had developed an ingenious way of looking at the world.

Penelope Boston, who assisted with the AGU conference, thinks the poetic name for the theory "may have been a strategic mistake." But Lovelock's daring view might have gone unnoticed without such language. If you have a big concept, you ought to have a name equal to it, advised the novelist William Golding, a friend and neighbor of Lovelock. And as Lovelock has joked, Gaia is "a more convenient term than biological cybernetic system with homeostatic tendencies."

The Gaia hypothesis has also started filtering into the mainstream in the guise of "Earth systems science." This new discipline, which spans a range of sciences, stresses the feedback mechanisms among the oceans, atmosphere, climate, geological systems, and biosphere. The need to understand global climate change has pushed the approach into the vanguard of international research. Walter Rosen, a biologist with the National Academy of Sciences, says global change "may be the ultimate test of the Gaia hypothesis."

Some geochemists resist the hypothesis since it contradicts their explanations of the Earth's atmospheric phenomena. Harvard University geochemist Heinrich Holland, a staunch opponent of the Gaia hypothesis, says, "There's very little scientific basis for the thing. Many of us are starting to feel we understand the geochemistry of the atmosphere and the oceans. We feel that Jim has gotten ahold of one piece, and that he is overemphasizing the biosphere vastly." Holland thinks Lovelock is ignoring the importance of the hydrologic cycle and volcanoes that spew out carbon dioxide and sulfur.

Atmospheric chemist James Kasting of Pennsylvania State University goes one step farther, arguing that Earth's temperature can remain stable and habitable for life even without biological influences. In the February 1988 issue of *Scientific American*, he and colleagues describe a process by which a geochemical feedback system could take carbon dioxide from the atmosphere and cool the planet's temperature. Kasting agrees with Lovelock that phytoplankton and blue-green algae remove carbon from the atmosphere and ocean and make calcium carbonate, but he counters that calcium ions in seawater also could combine with carbon inorganically to produce the same material.

Some scientists find aspects of the Gaia hypothesis useful in their research, however. Lee Kump, a geochemist at Pennsylvania State University, argues that geochemical models of atmospheric oxygen are missing some fundamental control processes. He refers to the widely cited "modified Blag model," which calculates changes in atmospheric oxygen through time. It indicates that oxygen levels should have fluctuated wildly in the past 600 million years. "This would have been impossible without dramatic consequences, such as the end of breathing organisms or disastrous forest fires," Kump says. "I'm convinced that the biosphere is playing a major role in the Earth's chemistry and climate, and I've been taken along this path by Lovelock."

A New Paradigm

Some scientists who have started listening to Lovelock think Gaia is less important as a concrete theory than as a new paradigm. "I see it not as a completely worked-out hypothesis but as a generator of new hypotheses," says Tyler Volk, an oceanographer at New York University.

Moreover, the Gaia idea is encouraging interdisciplinary study. In this respect it has had a major effect on one of the most ambitious international science programs, according to Walter Rosen at the National Academy of Sciences. In 1986 the International Council of Scientific Unions launched the decade-long International Geosphere-Biosphere Program (IGBP) to find out more about global climate change. "It's really pretty revolutionary. You have oceanographers talking to biologists, atmospheric chemists talking to microbiologists, scientists crossing all sorts of interdisciplinary boundaries," Rosen says. "I am convinced that Gaia influenced IGBP by putting that little switch in everyone's brain that leads to questions of the type 'What is the global biosphere's role?'" notes Volk.

Since the AGU conference, a number of scientists have started looking at how the biosphere may affect the environment. David Schwartzman, a geologist at Howard University, is studying the role of soil microbes in accelerating the absorption of atmospheric carbon dioxide through weathering. Gaian concepts have stimulated much of the work that Ralph Cicerone of the National Center for Atmospheric Research conducts on trace gases, particularly methane. And Lee Kump is examining how forest fires may regulate atmospheric oxygen levels. Moreover, at least half a dozen courses on the Gaia theory are being taught at universities around the country. Even some detractors such as James Kasting of Penn State are offering classes on the topic, because it makes students think about how Earth might work.

It's still too early to say what will happen to the Gaia hypothesis. Lovelock's collaborator, Lynn Margulis, thinks that feedback mechanisms will be discovered to prove the idea, although the work may take years. In the meantime, she points out that no one has yet uncovered mechanisms to prove Darwinian evolution either. And scientists like Lee Kump don't mind that the Gaia hypothesis isn't producing immediate breakthroughs. "Many of the big revolutions in science have been spawned by what were considered at the time to be crazy ideas—continental drift, for example. Lovelock is making us take another look at how the Earth works."

Gaia's Critics

In 1991 the debate continues. While some scientists (notably, earth systems scientists) have wholeheartedly begun the search for Gaian mechanisms, others are still staunchly opposed to the very notion of Gaia. Lovelock himself mentioned two of the most prominent Gaia critics: the biologists W. Ford Doolittle and Richard Dawkins. His Daisyworld model, in fact, was a response to their teleological criticisms.

*Citing the mechanism of natural selection but using arguments too technical to introduce here, Dawkins and Doolittle remain unconvinced by Daisyworld. The bibliography provides references for finding their critical views, and Ford Doolittle presents the crux of this argument in chapter 17. Additionally, in his initial critique of Gaia (*Coevolution Quarterly*, spring 1981), Doolittle makes an argument that has, in the last few decades, been taken very seriously in the field of astrophysics. Whether or not one finds it compelling in the realm of biology, scientists today need to be familiar with this so-called Anthropic Principle. (Doolittle's work is reprinted here with permission of Coevolution Quarterly/Whole Earth Review.)*

W. Ford Doolittle

I do not doubt that some of the feedback loops which Lovelock claims exist do exist, but I do doubt that they were created by natural selection, or that they are anything but accidental. Methane production may now balance oxygen production nicely, but it is not written into the genetic codes of either oxygen producers or methane producers that this should be so, and either could easily get out of hand. Accidental balances are fragile things, and their maintenance depends upon chance.

If the fitness of the terrestrial environment is accidental, then is Lovelock not right in saying that for life to have survived to reach the stage of self-awareness "is as unlikely as to survive unscathed a drive blindfold through rush-hour traffic"? I think he is right; the prolonged survival of life is an event of extraordinarily low probability. It is however an event which is a prerequisite for the existence of Jim Lovelock and thus for the formulation of the Gaia hypothesis, and I think it is therefore logically fallacious to assume that any explanation other than chance is required. Can we not assume that there is an immense number of planets on which life independently arose and then through some global catastrophe was extinguished before the evolution of self-awareness? And should we not expect that on those few planets on which intelligent beings arose, because such a catastrophe by chance did not occur, someone would seek to explain his own survival by the invocation of some protective device such as Gaia? Surely if a large enough number of blindfold drivers launched themselves into rush-hour traffic, one would survive, and surely he, unaware of the existence of his less fortunate colleagues, would suggest that something other than good luck was on his side.

Cosmologists seem to have grappled with a similar problem. B. J. Carr and M. S. Rees pointed out (in a 1979 issue of *Nature*, 278: 605–612) that "the possibility of life as we know it evolving in the Universe depends on the values of a few basic physical constants, and is in some respects remarkably sensitive to their numerical values." They do not assume, however, that life has manipulated these constants (which a priori could have many other values) for its own ends, because there is a simpler explanation. Only the known values are potentially observable, because other values would not have permitted the evolutionary development of observers. Similarly, only a world which behaved as if Gaia did exist is observable, because only such a world can produce observers.

Does it matter if Lovelock is wrong, and the apparent stability of the biosphere reflects not the operation of a global cybernetic system created by natural selection to maintain conditions which are optimal for life, but sheer good luck? Yes it does, because Gaia, if she exists, has built-in corrective mechanisms and buffering systems which will protect the biosphere against many potential threats to its survival. An accidental system, although it may have some accidentally-developed buffering capacity, is inherently fragile and cannot evolve in the adaptive way Gaia could (if she were real) in response to new threats.

3 *Philosophical Musings*

"Gaia may turn out to be the first religion to have a testable scientific theory embedded in it," speculates James Lovelock. Religion or no, Gaia has certainly spawned a good deal of spiritual musings. Here are several samples, drawn from the writings of Lewis Thomas, Gary Snyder, Joseph Campbell with Bill Moyers, and Lovelock himself.

The World's Biggest Membrane

Lewis Thomas views the Earth's atmosphere in a Gaian sort of way. He calls the atmosphere "the world's biggest membrane" in his award-winning book The Lives of a Cell, *published in 1974. Although trained in medicine (and a past chairman of the Department of Pathology at Yale University), Lewis Thomas is best known for his translation of medical and biological knowledge into beautifully crafted essays that are at once factual, personal, and philosophical. Excerpts here from "The World's Biggest Membrane" (copyright 1971 by The Massachusetts Medical Society) are reprinted by permission of the author and Viking Penguin, a division of Penguin Books USA Inc.*

Lewis Thomas

Viewed from the distance of the moon, the astonishing thing about the earth, catching the breath, is that it is alive. The photographs show the dry, pounded surface of the moon in the foreground, dead as an old bone. Aloft, floating free beneath the moist, gleaming membrane of bright blue sky, is the rising earth, the only exuberant thing in this part of the cosmos. If you could look long enough, you would see the swirling of the great drifts of white cloud, covering and uncovering the half-hidden masses of land. If you had been looking for a very long, geologic time, you could have seen the continents themselves in motion, drifting apart on their crustal plates, held afloat by the fire beneath. It has the organized, self-contained look of a live creature, full of information, marvelously skilled in handling the sun.

It takes a membrane to make sense out of disorder in biology. You have to be able to catch energy and hold it, storing precisely the needed amount and releasing it in measured shares. A cell does this, and so do the organelles inside. To stay alive, you have to be able to hold out against equilibrium, maintain imbalance, bank against entropy, and you can only transact business with membranes in our kind of world.

"Viewed from the distance of the moon, the astonishing thing about the earth, catching the breath, is that it is alive."—Lewis Thomas

When the earth came alive it began constructing its own membrane, for the general purpose of editing the sun. Originally, there was nothing to shield out ultraviolet radiation except the water itself. The first thin atmosphere came entirely from the degassing of the earth as it cooled, and there was only a vanishingly small trace of oxygen in it. The formation of oxygen had to await the emergence of photosynthetic cells, and these were required to live in an environment with sufficient visible light for photosynthesis but shielded at the same time against lethal ultraviolet. The green cells must therefore have been about ten meters below the surface of water, probably in pools and ponds shallow enough to lack strong convection currents.

It is another illustration of our fantastic luck that oxygen filters out the very bands of ultraviolet light that are most devastating for nucleic acids and proteins, while allowing full penetration of the visible light needed for photosynthesis. If it had not been for this semipermeability, we could never have come along. Now we are protected against lethal ultraviolet rays by a narrow rim of ozone, thirty miles out. We are safe, well ventilated, and incubated, provided we can avoid technologies that might fiddle with that ozone, or shift the levels of carbon-dioxide.

It is hard to feel affection for something as totally impersonal as the atmosphere, and yet there it is, as much a part and product of life as wine or bread. For sheer size and perfection of function, it is far and away the grandest product of collaboration in all of nature. It breathes for us, and it does another thing for our pleasure. Each day, millions of meteorites fall against the outer limits of the membrane and are burned to nothing by the friction. Without this shelter, our surface would long since have become the pounded powder of the moon. Even though our receptors are not sensitive enough to hear it, there is comfort in knowing that the sound is there overhead, like the random noise of rain on the roof at night.

By the time he assembled his 1983 collection of essays, Late Night Thoughts on Listening to Mahler's Ninth Symphony, *Lewis Thomas had adopted Gaia as more than a metaphor. This excerpt from "The Corner of the Eye" (copyright 1981 by Lewis Thomas) is reprinted by permission of the author and Viking Penguin, a division of Penguin Books USA Inc.*

The overwhelming astonishment, the queerest structure we know about so far in the whole universe, the greatest of all cosmological scientific puzzles, confounding all our efforts to comprehend it, is the earth. We are only now beginning to appreciate how strange and splendid it is, how it catches the breath, the loveliest object afloat around the sun, enclosed in its own blue bubble of atmosphere, manufacturing and breathing its own oxygen, fixing its own nitrogen from the air into its own soil, generating its own weather at the surface of its rain forests, constructing its own carapace from living parts: chalk cliffs, coral reefs, old fossils from earlier forms of life now covered by layers of new life meshed together around the globe, Troy upon Troy.

Seen from the right distance, from the corner of the eye of an extraterrestrial visitor, it must surely seem a single creature, clinging to the round warm stone, turning in the sun.

Songs for Gaia

Gary Snyder is more than a Pulitzer-prize-winning poet. He is a spiritual leader of the environmental movement, particularly those elements devoted to the concept of bioregionalism: living simply, lightly, and joyfully within the bounds set by one's home ecosystem. His 1983 collection of poems, Axe Handles, *offers an entire chapter extolling Gaia. Here are three selections (copyright 1983 by Gary Snyder and reprinted by permission of North Point Press).*

Gary Snyder

across salt marshes north of
San Francisco Bay
cloud soft grays
blues little fuzzies
illusion structures—pale blue of the edge,
    sky behind,

hawk dipping and circling
over salt marsh

ah, this slow-paced
system of systems, whirling and turning

a five-thousand-year span
    about all that a human can figure,

grasshopper man in his car driving through.

As the crickets' soft autumn hum
            is to us,
      so are we to the trees

          as are they

   to the rocks and the hills

Deep blue sea baby,
Deep blue sea.
                    Ge, Gaia
Seed syllable, "ah!"

Whirl of the white clouds over blue-green land and seas
        blue-green of bios   bow—curve—

Chuang-tzu says the Great Bird looking down,
        all he sees is
                    blue . . .

Sand hills. blue of the land, green of the sky.
        looking outward
        half-moon in cloud;

Red soil—blue sky—white cloud—grainy granite,
                                and
Twenty thousand mountain miles of manzanita.
    Some beautiful tiny manzanita
    I saw a single, perfect, lovely
            manzanita

                                Ha.

The Power of Myth—and Science

In the enormously popular television series The Power of Myth *(which in book form is on the* New York Times *best-seller list as I write), Bill Moyers converses with expert Joseph Campbell. Following a discussion about the religious strife in Beirut, Campbell declares, "We need myths that will identify the individual not with his local group but with the planet." Later an exchange about Gaia emerges. The excerpt is drawn from* The Power of Myth: Joseph Campbell with Bill Moyers, *edited by Betty Sue Flowers (copyright 1988 by Apostrophe S Productions, Inc., and Alfred van der Marck Editions and reprinted by permission of Doubleday).*

Joseph Campbell with Bill Moyers

Moyers: Scientists are beginning to talk quite openly about the Gaia principle.

Campbell: There you are, the whole planet as an organism.

Moyers: Mother Earth. Will new myths come from this image?

Campbell: Well, something might. You can't predict what a myth is going to be any more than you can predict what you're going to dream tonight. Myths and dreams come from the same place. They come from realizations of some kind that have then to find expression in symbolic

The Dream (Henri Rousseau, 1910).

form. And the only myth that is going to be worth thinking about in the immediate future is one that is talking about the planet, not the city, not these people, but the planet, and everybody on it. That's my main thought for what the future myth is going to be.

And what it will have to deal with will be exactly what all myths have dealt with—the maturation of the individual, from dependency through adulthood, through maturity, and then to the exit; and then how to relate to this society and how to relate this society to the world of nature and the cosmos. That's what myths have all talked about, and what this one's got to talk about. But the society that it's got to talk about is the society of the planet. And until that gets going, you don't have anything.

Moyers: There is that wonderful photograph of the Earth seen from space and it's very small yet, at the same time, it's very grand.

Campbell: When you see the earth from the moon, you don't see any division there of nations or states. This might be the symbol, really, for the new mythology to come. That is the country that we are going to be celebrating. And those are the people that we are one with.

Science and the Sacred

This excerpt is drawn from The Ages of Gaia *by James Lovelock (copyright 1988 by The Commonwealth Fund Book Program of Memorial Sloan-Kettering Cancer Center). It is reprinted here with permission of the author and W. W. Norton & Company, Inc.*

James Lovelock

When I first saw Gaia in my mind I felt as an astronaut must have done as he stood on the Moon, gazing back at our home, the Earth. Thinking of the Earth as alive makes it seem, on happy days, in the right places, as if the whole planet were celebrating a sacred ceremony. Being on the Earth brings that same special feeling of comfort that attaches to the celebration of any religion when it is seemly and when one is fit to receive.

Many, I suspect, have trodden this same path through the mind. Those millions of Christians who make a special place in their hearts for the Virgin Mary possibly respond as I do. The concept of Jahweh as remote, all-powerful, all-seeing is either frightening or unapproachable. Even the sense of presence of a more contemporary God, a still, small voice within, may not be enough for those who need to communicate with someone outside. Mary is close and can be talked to. She is believable and manageable. It could be that the importance of the Virgin Mary in faith is something of this kind, but there may be more to it.

What if Mary is another name for Gaia? Then her capacity for virgin birth is no miracle or parthenogenetic aberration; it is a role of Gaia since life began. Immortals do not need to reproduce an image of themselves; it is enough to renew continuously the life that constitutes them. Any living organism a quarter as old as the Universe itself and still full of vigor is as near immortal as we ever need to know. She is of this Universe and, conceivably, a part of God. On Earth she is the source of life everlasting and is alive now; she gave birth to humankind and we are a part of her.

The belief that the Earth is alive and to be revered is still held in such remote places as the west of Ireland and the rural parts of some Latin countries. In these places, the shrines to the Virgin Mary seem to mean more, and to attract more loving care and attention, than does the church itself. The shrines are almost always in the open, exposed to the rain and to the sun, and surrounded by carefully tended flowers and shrubs. I cannot help but think that these country folk are worshipping something more than the Christian maiden.

There is little time left to prevent the destruction of the forests of the humid tropics with consequences far-reaching both for Gaia and for humans. The country folk, who are destroying their own forests, are often Christians and venerate the Holy Virgin Mary. If their hearts and minds could be moved to see in her the embodiment of Gaia, then they might become aware that the victim of their destruction was indeed the Mother of humankind and the source of everlasting life.

When that great and good man Pope John Paul travels around the world, he, in an act of great humility and respect for the Mother or Father Land, bends down and kisses the airport tarmac. I sometimes imagine him walking those few steps beyond the dead concrete to kiss the living grass; part of our true Mother and of ourselves.

For me, Gaia is a religious as well as a scientific concept, and in both spheres it is manageable. The life of a scientist who is a natural philosopher can be deeply religious. Curiosity is an intimate part of the process of loving. Being curious and getting to know the natural world leads to a loving relationship with it. It can be so deep that it cannot be articulated, but it is nonetheless good science. Creative scientists, when asked how they came upon some great discovery, frequently state, "I knew it intuitively, but it took several years work to prove it to my colleagues." Compare that statement with this one by William James, the nineteenth-century philosopher and psychologist, in *The Varieties of Religious Experience*:

> The truth is that in the metaphysical and religious sphere, articulate reasons are cogent for us only when our inarticulate feelings of reality have already been impressed in favor of the same conclusion. Then, indeed, our intuitions and our reason work together, and great world ruling systems, like that of the Buddhist or of the Catholic philosophy, may grow up. Our impulsive belief is here always what sets up the original body of truth, and our articulately verbalized philosophy is but a showy translation into formulas. The unreasoned and immediate assurance is the deep thing in us, the reasoned argument is but a surface exhibition. Instinct leads, intelligence does but follow.

"Interior of the Primeval Forest on the Amazons" (Henry Walter Bates, ca. 1850).

This was the way of the natural philosophers in the eighteenth century and is still the way of many scientists today. Science can embrace the notion of the Earth as a superorganism and can still wonder about the meaning of the Universe.

"Sagan Urges Clerics to Join in an Effort to Save the Globe" reads the title of a New York Times *story (16 January 1990). Reporter Peter Steinfels tells of an initiative taken by Carl Sagan, who, you may recall, is the editor of the journal* Icarus *and who published Lovelock's first paper on the Gaia hypothesis. Carl Sagan was also one of a score of scientists who, in the early 1980s, alerted the world to the danger posed by "nuclear winter" (a cold, dark period) in the aftermath of nuclear war. Here is an excerpt, copyright 1990 by The New York Times Company and reprinted by permission.*

Peter Steinfels

Carl Sagan, a professor of astronomy and director of the Laboratory for Planetary Studies at Cornell University, appealed yesterday for religion and science to join hands in preserving the global environment. He was joined in his appeal by 22 well-known scientists. "Efforts to safeguard and cherish the environment need to be infused with a vision of the sacred," the scientists said.

The statement listed the dangers of global warming, the depletion of the ozone layer, the extinction of plant and animal species, the destruction of rain forests and the threat of nuclear war. "Problems of such magnitude," the scientists said, "and solutions demanding so broad a perspective, must be recognized from the outset as having a religious as well as a scientific dimension."

Dr. Sagan made the appeal public in Moscow on the first day of a conference on the environment and economic development. Last week, before leaving for Moscow, Dr. Sagan said the appeal was significant because of the historic antagonisms between science and religion. "I am personally skeptical about many aspects of revealed religion," Dr. Sagan said. "But I am sure of the awe and reverence that the meticulously balanced nature of the global environment elicits in me."

About a hundred religious leaders have signed a statement welcoming the appeal as "a unique moment and opportunity in the relationship of science and religion." The Rev. James Parks Morton, Dean of the Episcopal Cathedral of St. John the Divine in New York and an organizer of the Moscow conference, said in Moscow that scientists and religious leaders would meet immediately to discuss "practical next steps."

Physicist Freeman Dyson and biologist Stephen Jay Gould were among the scientists who joined Sagan in making the appeal.

Chief Seattle

Will you teach your children what we have taught our children?
That the earth is our mother?

II *Merged Beings*

Little Fly,
Thy summer's play
My thoughtless hand
Has brush'd away

Am not I
A fly like thee?
Or art not thou
A man like me?

—William Blake

We are beginning to see the biosphere not only as a continual struggle favoring the most vicious organisms but also as an endless dance of diversifying life forms, where partners triumph.

—Lynn Margulis and Dorion Sagan

4 *Lynn Margulis: Vindicated Heretic*

Freelance writer Jeanne McDermott produced a profile of Lynn Margulis for the August 1989 issue of Smithsonian *magazine (copyright 1989 by Jeanne McDermott). Here are excerpts from "A Biologist Whose Heresy Redraws Earth's Tree of Life," reprinted by permission of the author and Smithsonian magazine.*

Jeanne McDermott

The first thing everyone notices about Lynn Margulis is that it's impossible to keep up with her. She dresses for comfort but her mind obviously thrives on work. She wears corduroy jeans and a colorful Peruvian sweater, a gift from one of her four children. She rises at 5:30 and bicycles to her modest book-lined office at the University of Massachusetts at Amherst before the rest of the campus awakens. "That's when I get my work done," she says. From 9 to 5, the quiet dissolves into a whirlwind of activity. Friends and colleagues have a shorthand way of acknowledging her prodigious energy. "Oh, you know Lynn," they say.

As the director of a $100,000-a-year research lab, she taught two or three courses at Boston University every semester for 22 years, and is continuing this work at the University of Massachusetts. She has authored more than 130 scientific articles and 7 books. She rarely sits still. Her speech is nonstop and, in the jargon of her profession, filled with references to DNA homologies, microtubules and antitubulin probes. She interrupts and digresses constantly because one idea triggers an avalanche of others. She likes to make daring statements such as, "The nervous system is explicable in terms of microbiology." "That's propaganda" is her way of dismissing dogma, which she hates. She is restless, passionately curious, irreverent, sassy and very sharp.

One colleague calls her the most gifted theoretical biologist of her generation. Another, of the century. "Her mind keeps shooting off sparks," says Peter Raven, director of the Missouri Botanical Garden and a MacArthur Fellow. "Some critics say she's off in left field. To me, she's one of the most exciting, original thinkers in the whole field of biology."

Margulis is an authority on the microcosmos. She likely knows more than anyone else about the role of microorganisms in the past 3.5 billion years of evolution. While most American biologists emphasize the role of competition in evolution, Margulis stresses symbiosis. For example, she views each human cell as a "sophisticated aggregate of evolving microbial life."

Lynn Margulis

Challenging a Worldview

As children, we gradually learn how the world works—china cups break when they fall, hot coffee burns, the sun sets at night. As any summer visitor to northern Alaska can attest, it can be very upsetting when the world does not work as expected. Scientists, likewise, build their own notions of the way the world works, which MIT philosopher Thomas Kuhn calls scientific paradigms. Periodically, but not very often, a scientist will make a discovery so outlandish that it does not fit into any paradigm, leading to a sort of intellectual earthquake or, in Kuhn's language, a paradigm shift. Margulis is one of the few living scientists who has shifted a paradigm.

"When I was an undergraduate, two theories were held up for ridicule, to show how farfetched scientific theories can get," says William Culberson, professor of botany at Duke University. "One was the theory of continental drift and the other was the symbiotic theory of the origin of the cell. Neither is laughed at today. The reason that the symbiotic theory is taken seriously is Margulis. She's changed the way we look at the cell."

Yet, Margulis has spent much of her career on the margins of respectability, battling the scientific community's lack of familiarity with the more than 200,000 known species of microbes on Earth, most of which do nothing that directly harms or helps the human race. "Microbiology was, historically, a practical art, not a science," she says. "You know, kill the germs and save the food." Only in the past ten years have science textbooks begun to reflect her views. Ironically, she herself has become the new authority. "It's worrisome," she says. "Depressing. Authorities change. The experience doesn't. When science is taught by reading a textbook, you open the door to dogma." Her students get their feet wet.

Despite her success, Margulis's work remains controversial. Hers is not the kind of work with which the scientific community can simply agree to disagree. "It's a question of changing your religion," she says. Academia rewards its brightest stars with a specially funded teaching position called a named chair. A few years ago, Margulis was on the verge of being appointed to a named chair at a major university but was not offered the position, though the possibility still remains. The antagonism stems, in part, from Margulis's collaboration with British chemist James Lovelock on Gaia, the hypothesis that the Earth acts as a self-regulating, self-maintaining system. Gaia's most vocal supporters are ecoactivists, church groups and science fiction writers. To some establishment scientists, these countercultural associations make the ideas behind Gaia suspect. Ironically, Margulis is hard on Gaia's popular supporters. "Lynn is ferocious about going after mysticism," says Stewart Brand, founder of the *Whole Earth Catalog*. "New Age types are drawn to her and then she busts them high, low and center for being softheaded."

"All of us are walking communities"

In a sense, Margulis challenges the American myth of the rugged individual—alone, self-contained and able to survive. "Our concept of the individual is totally warped," she says. "All of us are walking communities. Every plant and animal on Earth today is a symbiont, living in close contact with others."

Consider one species of desert termite. Living in its hindgut are millions of single-celled, lemon-shaped organisms called *Trichonympha ampla*. Attached to the surface of one *T. ampla* live thousands of whiplike bacteria known as spirochetes *(pronounced spy'-row-keets)*. Inside live still other kinds of bacteria. If not for these microbial symbionts (in some wood-eating insects, the symbionts are too numerous to count), the termite, unable to digest wood, would starve.

But, the termite itself is only one element in a planetary set of interlocking, mutual interactions—which Lovelock's neighbor, novelist William Golding, dubbed Gaia, for the Greek goddess of the Earth. After digesting wood, the termite expels the gas methane into the air. (In fact, the world's species of termites, cows, elephants and other animals harboring methane-producing bacteria account for a significant portion of Earth's atmospheric methane.) Methane performs the vital task of regulating the amount of oxygen in Earth's atmosphere. If there were too much oxygen, fires would burn continuously; too little, and animals, plants and many other live beings would suffocate. Earth's atmospheric oxygen is maintained, altered and regulated by the breathing activities of living creatures, such as those of the methane-makers in the microcosmos. Life does not passively "adapt." Rather, it actively, though "unknowingly," modifies its own environment.

"Gaia is more a point of view than a theory," says Margulis. "It is a manifestation of the organization of the planet." That organization resembles those hollow Russian dolls that nest one inside another. "For example, some bacteria in the hindgut of a termite cannot survive outside that microbial community," explains Gail Fleischaker, Boston University philosopher of science and a former graduate student of Margulis's. "The community of termites, in turn, requires a larger ecological nest. And so it expands. You will never find life in isolation. Life, if it exists at all, is globe covering."

Although Margulis provided the "biological ammunition" for Gaia and remains its staunch advocate, she does little work on it directly. "I've concentrated all my life on the cell," she says. The ideas that she has championed were once "too fantastic for mention in polite biological society," as one scientific observer described them in the 1920s. As recently as twenty years ago, these ideas were so much at odds with the established point of view that, according to another observer, they "could not be discussed at respectable scientific meetings."

Although aspects of the symbiotic theory of cell evolution still provoke hostility, the theory is now taught to high school students. "This quiet revolution in microbiological thought is primarily due to the insight and enthusiasm of Lynn Margulis," states Yale ecologist G. Evelyn Hutchinson. "Hers is one of the most constructively speculative minds, immensely learned, highly imaginative and occasionally a little naughty."

Creating a Career

There is little separation between Margulis's professional and personal lives. You're just as likely to find her children in the lab as at her home. She drops her grandson Tonio at preschool on the way to hear a student defend his thesis. There are times when you wonder if she runs a family business.

Dorion Sagan, her eldest son from her first marriage, coauthored the popular book *Microcosmos* along with five other scientific books. Her daughter-in-law Christie Lyons has illustrated most of them. Students say that they are treated like extended family; she shares the benefits of her prestige, insisting, as other scientists of her stature rarely do, that they are invited as well when she speaks at exclusive scientific meetings. "That's the way it should be," says Lovelock. "That's something consonant with the theories we've worked on. We're ecumenical people."

Unlike most authorities on microbes, Margulis has never taken a microbiology course. She was born in 1938 and grew up in Chicago, the eldest of four sisters. Intellectually precocious, she started the University of Chicago's undergraduate program at 14. It was the Great Books era at Chicago. Chancellor Robert Hutchins had instituted a curriculum in which students read original works, following the development of ideas rather than academic disciplines, an intellectual approach that Margulis embraces to this day. In her second year, she took a natural science course that examined the question: What is heredity? Watson and Crick had just discovered DNA, and she became fascinated. She also met a fellow student whom she later married, astronomer Carl Sagan.

Dorion Sagan laughingly refers to his parents as his Earth mother and Space father. Indeed, while Carl Sagan emphasizes the cosmos of the stars, Margulis stresses the microcosmos on Earth. Where Sagan speculates about extraterrestrial life, Margulis insists that life be seen as a planetary phenomenon whose Earth-bound limits remain unexplored.

After graduation, Margulis and Sagan moved to Madison where she pursued a master's degree in zoology and genetics at the University of Wisconsin. There she studied with cell biologist Hans Ris. They discussed puzzling patterns of heredity that could not be explained by the conventional assumption that the genes in the cell's nucleus solely determined the genetic make-up of its offspring. In 1960, Margulis followed Sagan to Berkeley. For her dissertation she wanted to look for genes in the cell's cytoplasm, the area outside the nucleus; her professors tried to talk her out of it.

By 1963, Ris and W. S. Plaut—another of Margulis's professors at Wisconsin—published a photograph showing that DNA resided in the cytoplasm, specifically in the chloroplasts, tiny organelles found in plant cells. (Organelles are the visible bodies and structures inside a cell.) The presence of DNA outside the nucleus baffled the biological community. But not Margulis. "In Ris's classes, I had read the cell biology and genetics literature of the late 19th and early 20th centuries." She remembered the "crackpot" idea that the genes in the cytoplasm and the genes in the nucleus had different origins in evolution.

The first creatures to evolve happened to be the last creatures that scientists discovered. It wasn't until the advent of good microscopes in the 19th century that the scientific community began to appreciate the diversity of the microbial world. No one knew how to classify microbes. Zoologists called the moving little things animals. Botanists called the green little things plants. But neither classification made sense. It turns out that the most profound difference between living creatures lies not in their color or their locomotion but in the cells of which they are made.

There are only two types of cells on Earth. One type has a nucleus. The other type does not. Animals and plants are made of nucleated cells. Most microbes are made of nonnucleated cells. At the turn of the century, Russian biologist K. S. Mereschkovsky proposed the idea that plant and animal cells evolved from symbioses between bacterial cells. He based his theory on the similar appearances and behaviors of free-living bacteria and the organelles of plant and animal cells. Margulis saw that the advances in electron microscopy and molecular biology vastly expanded the quality and quantity of evidence.

The years from 1963 to 1967 were not easy. Margulis got her PhD at 26 and moved to Boston; she and Sagan ended their marriage. Without a formal academic position, she continued to build a case for the symbiotic theory. A big break came when she saw a list, drawn up by British crystallographer J. D. Bernal, of the most important unresolved biological questions of the times. One was the origin of the nucleated, or eukaryotic, cell. She immediately sent Bernal a three-page reply and, having mimeographed it, mailed copies to scientists interested in cell evolution. Her solution was the theory that cells with nuclei evolved from the merger of two or more different bacterial cells lacking nuclei. The responses to her letter resembled a Rorschach test. One scientist corrected her grammar. Another ordered her out of his field. "Bernal said I'd solved the problem," she says.

By 1970, she was an associate professor at Boston University and was married to Thomas N. Margulis; she was the mother of two more children and author of the first comprehensive account of her ideas, *The Origin of Eukaryotic Cells*. One reviewer wrote, "Readers will find this book sprawling, stimulating, irritating and challenging but they will have difficulty ignoring it." Another wrote that "it had to be a young scientist and a woman who dared to challenge the scientific establishment."

As one former student puts it, the scientific community tried to shut her up by attempting to prove her wrong and by denying her grant support. Ironically, she turned out to be right, no doubt a factor in some of the controversy she generates. "She's made people think more than any other figure in contemporary biology, and for many men it is particularly galling that this has been done by a woman," says Culberson. By 1981, when Margulis published *Symbiosis in Cell Evolution*, a new version of the thesis of her first book, the scientific establishment had finally accepted the notion that mitochondria and chloroplasts evolved symbiotically. "I always thought that everyone would catch up," Margulis says wryly.

With acceptance came "semifame," according to Betsey Dyer, a former graduate student of Margulis's and now professor of biology at Wheaton College in Norton, Massachusetts. Grant money that had been impossible to obtain came in with more ease. Margulis could no longer answer her own phone. The lab tripled in size. Then, in 1983, she received the honor that is second only to the Nobel for an American scientist: she was elected to the National Academy of Sciences. "They called out of the blue. I was totally shocked. Amazed," she says. The certificate hangs on her office wall, along with pictures of the children and one taken of her scavenging mudflats in Baja California for spirochetes.

Mudflats, Microscopes, and Microtubules

As might be expected from someone who thrives on combating the status quo, Margulis has not been content to rest on her laurels. She is avidly gathering evidence to prove the last and by far the most controversial, most heretical chapter of the symbiotic theory. "We believe that the nucleated cell's internal transportation system, its ability to move within, is the contribution of another symbiotic merger with bacteria, this time with rapid whiplashing spirochetes," she writes.

Spirochetes are tiny, spiral-shaped, highly mobile bacteria, of which the most notorious is the one that causes syphilis. In mudflats along the coast of Baja California where conditions bear a resemblance to those found on Earth three billion years ago, Margulis discovered a spirochete that, she believes, is like the ancestors of the "tail"-bearing cells in the human body, such as the sperm cells and those that line the lungs or oviduct. She also believes that related spirochetes were the ancestors of

This highly magnified photograph (x 83,000) by a transmission electron microscope shows a group of sperm tails in cross-section during their maturation in the testis of a moth. Each contains an array of microtubules: 2 dead center, surrounded by a ring of 9 microtubule doublets, and beyond that another ring of 9 microtubule singlets pressed against the plasma membrane of the cell. The beating of the flagellum is accomplished by the microtubule doublets sliding past one another in a process fueled by ATP.

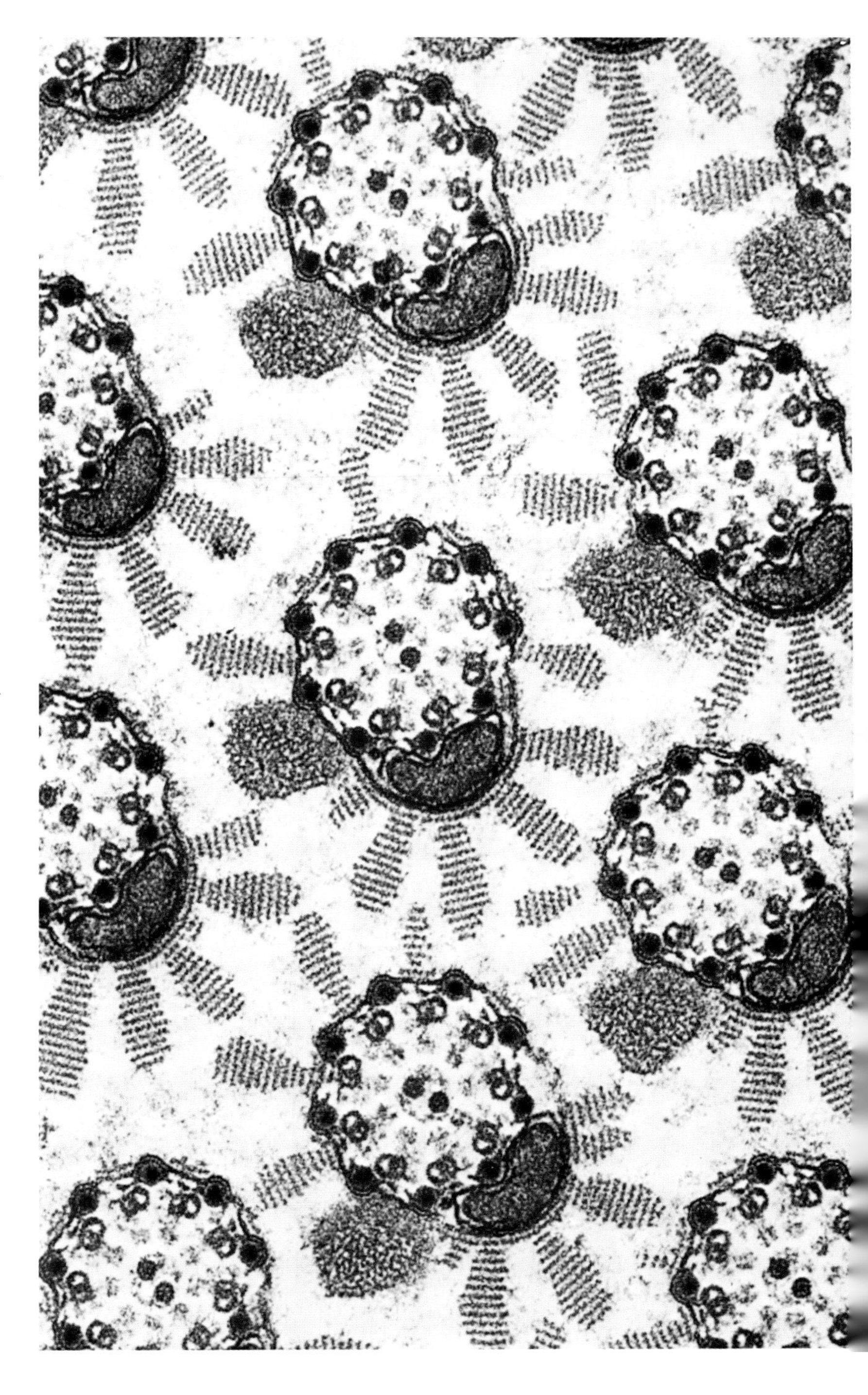

an entire intracellular transportation network, constructed of microtubules, which includes the tails and the mitotic apparatus responsible for chromosome movement in cell division.

This chapter in the symbiotic theory is considerably harder to prove. Unlike the ancestors of the mitochondria and the chloroplasts, early spirochetes left almost no tangible clues behind. Only in the past year has some tantalizing new evidence emerged. "The most important discovery of the century in cell biology is the finding made by David Luck and his colleague John Hall, at Rockefeller University, of kinetosomal DNA," says Margulis. She believes that DNA in the kinetosome, the structure from which sperm and all other cell tails grow, had at one time belonged to free-living spirochetes.

"Most people in microbiology know this hypothesis but can't take it seriously. There's no reasonable body of evidence. My attitude is wait and see, and I'm more sympathetic than many," says Peter Greenberg, microbiologist at the University of Iowa. Supporters say it misses the point to argue that Margulis may be wrong. They cite this work as an example of her excellence as a theoretician who pushes the frontiers of current thinking with provocative, testable hypotheses. Other scientists scoff. At a conference several years ago, a former Margulis student, University of North Texas biologist Stephen Fracek, caught "a lot of flak" when he presented some preliminary data. At a postconference party, one of the staunchest critics said that he could never support the theory, even if Margulis demonstrated amino acid homologies. "There is enormous hostility to the ideas behind the data," Margulis readily acknowledges. "Some people resist the concept that the tails of their own sperm evolved from free-living spirochetes."

Margulis, who believes "absolutely" that the case for spirochete ancestry will be proved, says that it will "change everything. Neurobiology, for example." Then she speculates, "You can reduce the study of the nervous system to physics and chemistry, but you're missing the microbiological step. It's as if you documented the changing surface of the Earth at urban sites, using Landsat *(satellite)* images, without knowing anything about people. In the nerve cell, the axons and the dendrites that make the physical connections that allow us to communicate are latter-day spirochetes. Nerve cells, having long ago discarded the rest of the spirochete body, use the fundamental motility system of spirochetes. Think of the nerve as coming from what had formerly been a bacterium, 'trying' but unable to rotate and swim. Thought involves motility and communication, the connection between remnant spirochetes. All I ask is that we compare human consciousness with spirochete ecology."

"History may prove she isn't always right," says G. Evelyn Hutchinson. "She often puts things in a dramatic way." But Margulis cares little for always being right. She cares about stripping the rungs off the evolutionary ladder, puncturing the anthropocentric view of life and encouraging her students to ask embarrassing questions. Her latest projects (coauthored with Dorion), aimed at the next generation of scientists, expose young minds to the biological riches of unseen worlds. They are called *The Garden of Microbial Delights* and *The Microcosmos Coloring Book*.

5 Microcosmos

Excerpts here are drawn in part from "Bacterial Bedfellows," an essay by Dorion Sagan and Lynn Margulis that appeared in the March 1987 issue of Natural History *(reprinted by permission of the authors and Natural History magazine). Excerpts are also drawn from* Microcosmos *by Lynn Margulis and Dorion Sagan (copyright 1986 by Lynn Margulis and Dorion Sagan and reprinted by permission of the authors and Summit Books, a division of Simon and Schuster, Inc.).*

Lynn Margulis,
Dorion Sagan

Human religion and mythology have always been full of fantastic combinations of creatures—the mermaids, sphinxes, centaurs, devils, vampires, werewolves, and seraphs that combine animal parts to make imaginary beings. Truth being stranger than fiction, biology has refined the intuitively pleasing idea with its discovery of the overwhelming probability of the reality of combined beings. We and all beings made of nucleated cells are probably composites, mergers of once different creatures.

Living corporations, some of the mergers began as hostile take-overs of one organism by another. But over hundreds of millions of years they became so coordinated, so interwoven that it took the electron microscope and the intricate techniques of biochemical analysis to penetrate the illusion that because the harmony of cell parts seemed perfect, it had always been so.

The Case of the Sick Amoebas

No one has lived long enough to witness the origin of species in the field. But in one case in the laboratory, a new variety of microbe evolved so quickly it was caught in the act. This event was witnessed and described by Kwang Jeon, a brilliant and keenly observant scientist in the zoology department of the University of Tennessee. The odyssey recorded by Jeon shows the dynamic we think responsible for the rapid evolution of cells with nuclei from bacteria about 1,500 million years ago. The story strongly demonstrates the inevitability of some kind of cooperation, called symbiosis, among organisms that are to live together and survive. It shows the thin line between evolutionary competition and cooperation. In the microcosm, guests and prisoners can be the same thing, and the deadliest enemies can become indispensable to survival.

Kwang Jeon had been raising and experimenting with amoebas for years when he welcomed a new batch to his laboratory. After putting the new batch into small bowls next to other amoebas gathered from all around the world, he noticed the spreading of a severe illness. Healthy amoebas grew round and granulated. They refused to eat and failed to reproduce. Bowl by bowl, more and more amoebas died. The few that grew and divided at all did so reluctantly, about once a month instead of once every other day.

When Jeon examined the dead and dying forms under the microscope, he noticed tiny spots inside the cells. On closer inspection he saw that about 100,000 rod-shaped bacteria, brought in by the new amoebas, were present in each amoeba. The rod bacteria had infected the rest of his collection. Yet the disease was not a total catastrophe. A small minority of infected amoebas survived the scourge. These "bacterized" amoebas were fragile organisms, oversensitive to heat, cold, and starvation. They were easily killed by antibiotics, which, while deadly to bacteria, did not harm his normal "nonbacterized" amoebas. A change was occurring. The two types of organisms, bacteria and amoebas, were becoming one.

For some five years, Jeon nurtured the infected amoebas back to health by selecting the tougher ones and letting the others die. Still infected, the amoebas began to divide again at the normal rate of once every other day. Reproductively speaking, they were as adapted as their uninfected ancestors. They were not rid of their bacteria—they all harbored "germs." But they were cured of their disease. Each recovered amoeba contained about 40,000 bacteria.

For their part, the bacteria had dramatically adjusted their destructive tendencies in order to live inside other living cells. Thus, from a violent confrontation emerged a new symbiotic organism: bacterized amoebas. Now, some ten years after the plague, the permanently infected amoebas are no longer sick, but alive and well and living in Knoxville, Tennessee.

The story does not end here. From friends, Jeon reclaimed some of the amoebas that he had sent off before the epidemic and which had never been exposed to the pathogenic bacteria. With a hooked glass needle, he then removed the nuclei from both infected and uninfected organisms and exchanged them. The infected amoebas with new nuclei lived on indefinitely. But the "clean" amoebas supplied with nuclei from cells that had been infected for years struggled for about four days and then died. It seemed as if the nuclei had become unable to cope with a "healthy" cell. Had they actually come to need their bacterial infection?

To find out, Jeon prepared another batch and mounted a rescue. Just a day or so before the bacterialess amoebas with their new nuclei would have died, he injected some of them with a few bacteria. The bacteria rapidly increased to the level of about 40,000 per cell, and the sick amoebas returned to health. A symbiotic habit had been formed; the bacteria were the "fix." Jeon's amoebas can be killed by penicillin, which binds to sites in the cell walls of the bacteria within them, destroying the interdependent population that is the cell. The pact between bacteria and amoebas has become so intimate and strong that death to one member of the alliance spells death for both.

Competition or Cooperation?

Jeon's bacteria showed that the only differences between organisms that kill or sicken, organisms that live together, and the indispensable components of organisms are differences of degree. Dangerous pathogens can become required organelles in less than a decade—very suddenly indeed considering the 3.5 billion years or so of biological evolution. Symbiosis leads abruptly to new species.

The amoeba experiments point out the fallacy of the idea that evolution works at all times for the "good of the individual." Just what is the "individual" after all? Is it the "single" amoeba with its internalized bacteria, or is it the "single" bacterium living in the cellular environment which is itself alive? Really, the individual is something abstract, a category, a conception. And nature has a tendency to evolve that which is beyond any narrow category or conception.

One such conception is the popular idea that evolution is a bloody struggle in which only the strong survive. "Survival of the fittest," a motto coined by the philosopher Herbert Spencer (1820–1903), was used by late-nineteenth century entrepreneurs to justify such mean practices as child labor, slave wages, and brutal working conditions. Warped to mean that only the most ruthless win out in the "struggle for existence," it also implied that exploitation, since it was natural, was morally acceptable.

Darwin would have been shocked at the misuse of his ideas. He used Spencer's phrase "survival of the fittest" to refer not to large muscles, predatory habits, or the master's whip but to leaving more offspring. Fit, in evolution, means fecund. The point is not so much the infliction of death, which is inevitable, as the propagation of life, which is not.

Competition in which the strong wins has been given a good deal more press than cooperation. But certain superficially weak organisms have survived in the long run by being part of collectives, while the so-called strong ones, never learning the trick of cooperation, have been dumped onto the scrap heap of evolutionary extinction. If symbiosis is as prevalent and as important in the history of life as it seems to be, we must rethink biology from the beginning.

Life on earth is not really a game in which some organisms beat others and win. It is what in the mathematical field of game theory is known as a nonzero-sum game. A zero-sum game is one like Ping-Pong or chess, where one player wins at the expense of his opponent's loss. An example of a nonzero-sum game is children playing house, or war: more than one player can win, and more than one side can lose.

Symbiosis and Cell Evolution

Kwang Jeon caught evolution in the act. What is more, the evolution of a new organism occurred by symbiosis, not by an accumulation of mutations. Furthermore, the new amoebas evolved not over millions of years but in eighteen months, which geologically speaking is instantaneous. Natural selection eliminated not competitors but competition itself. After the smoke had cleared, only symbionts—bacteria and amoebas that could work and live together—survived.

We should not be surprised. The deadliest parasites destroy not only their hosts and habitats but also their own chances for continued survival. *Consider the AIDS virus. So deadly and so recent a threat to humans, it seems to have worked out a compromise for coexistence with its long-time hosts, African monkeys. In contrast, the common cold virus, more a nuisance than a threat, even provokes symptoms, like coughing, that foster its propagation.*

Jeon's tale of two microbes hints at the answer to a major evolutionary puzzle. Of all the missing links in evolution, none is more profound than the gap between eukaryotes, cells with nuclei, and all bacteria, which lack nuclei. The difference between bacteria and any nucleated cells makes the difference between people and apes look negligible. Plant and animal cells have far more in common than do bacteria and nucleated cells. Cells with nuclei contain up to a thousand times more genetic material than their smaller relatives. This material is tightly coiled into chromosomes that are contained in a membrane-bounded nucleus.

Nucleated cells divide by a complex "dance of the chromosomes," during which the chromosomes pull the hereditary material to opposite ends of the cell and then divide. Bacterial cells simply split apart; they don't form chromosomes. They indulge in a wide range of metabolic variations, consume nitrogen and sulfur, produce methane, precipitate iron and manganese while breathing, and grow in boiling water and brine. Bacteria obtain their food and energy by using every sort of plant fiber and animal waste. If they did not, we would be living in a mounting heap of garbage.

A microscopic look at the waters of the earth 2,500 million years ago would have revealed flotillas of bobbing purple, blue-green, red, and yellow spheres: colonies of organisms crowding on rocks, gliding on water, or darting about with whipping tails. Shoals of bacterial cells waved with the currents, coating pebbles with brilliant hues. Bacterial spores blown by breezes showered the muddy terrain. Their genetic material, DNA and RNA, was not bound up; their genes were not packed into chromosomes wrapped by a nuclear membrane. They

Krishna as Navagunjara

reproduced asexually by growing to twice their size, replicating their single strand of DNA, and then dividing, with one copy of the DNA going to each offspring cell. Or a small cell containing a complete set of genetic material budded on the parent and then broke off. They also encased their DNA in spores that survived long periods of dryness, waiting to come alive when conditions became wetter or more generally favorable.

By 1,500 million years ago, the earth's modern surface and atmosphere were largely established and the bacteria flourished. Microbial life permeated the air, soil, and water, recycling gases and other compounds as they do today. From this low-lying milieu came new forms of life. A new kind of cell formed, larger and more complex than bacteria. This cell had circuitous channels of internal membrane, including one enveloping the nucleus. It had parts called mitochondria: dark bodies providing the cell surrounding them with energy derived from oxygen. Some would soon have plastids, chlorophyll-bearing packets capable of photosynthesis, suspended in their cytoplasm.

What brought about this new cell? As with other evolutionary puzzles, the solution to the mystery of the origin of the nucleated cell lies first in circumstantial evidence. History must be reconstructed from clues. If the ancestors of mitochondria were themselves bacteria without nuclei that raided and reproduced inside their hosts without killing them—in a fashion similar to Kwang Jeon's "dots"—an ancestral line of complex cells could have become established. There would be no record of transitional forms because the new entity would have evolved rapidly, the result of interspecies merger.

Mystery of Mitchondria

Imagine the ancestor of the mitochondria: a bacterial attacker, capable of breathing oxygen or even doing without it when necessary. Such microscopic predators still exist. *Bdellovibria* (the Greek "bdello" means leech; "vibrio" refers to their vibrating comma shape) burst asunder bacterial prey, eating them from the inside out. *Daptobacter* (the "gnawing bacterium") enters both the inner and the outer membranes of its victim's cell walls. Then it divides, again and again. The mitochondrial ancestor's original prey may have been a larger bacterium like modern-day *Thermoplasma*. The DNA of *Thermoplasma* is unlike that of other bacteria and similar to that of eukaryotes. This rugged bacterium can survive very hot and acidic water such as that found in the hot springs of Yellowstone National Park.

When they were first invaded, occupied hosts like *Thermoplasma* probably couldn't survive, and when they died, they took the invaders with them. Eventually, some of the prey evolved a tolerance for their predators, which then remained alive and well in the food-rich interior of their hosts. As they reproduced inside the invaded cells without causing harm, the predators gave up their independence and moved in for good. The two organisms thrived on internal leftovers—the products of each other's metabolism.

Invaded victims and tamed mitochondria recovered from the attack and have lived ever since, for 1,000 million years, in dynamic alliance. Because of the mitochondria, all earthly beings made of nucleated cells—which includes fungi, plants, animals, humans, and all organisms except bacteria—have remarkably similar metabolisms. *We eukaryotes all depend on oxygen processed by mitochondria for converting food into new tissues or into energy for performing work. Even green plants use oxygen for respiration, although the amount of oxygen they consume in converting the stored starches and sugars into plant tissues is less than the amount of oxygen they release as a waste product while "creating" these foods (through photosynthesis) in the first place.*

The presence of DNA in mitochondria helped tip off scientists to the possibility that these cells used to be free-living bacteria. When this DNA was examined, it was found to resemble the DNA in certain free-living bacteria far more than it resembled the DNA in the nucleus of the cell from which the mitochondria had come. Mitochondria have their own genes, their own reproductive timetable, and they often divide out of step with the rest of the cell. The bacteria that became mitochondria in our cells can be thought of as raiders that took over their hosts and formed cells with nuclei—cells ancestral to every plant and animal on this planet.

Roots of Photosynthesis

If we know where to go and how to look, we can see that these kinds of mergers are still occurring today. In a scene from a beautifully colorful silent film called *Intimate Strangers*, Oxford University botanist David C. Smith stands on a beach on the Brittany coast of France. Beneath his feet is what appears to be seaweed. But as Smith begins stepping on the spinachy green matter, it squirms straight down into the sand. Very soon all that's left is a cleared patch of beach. Where did it go?

Actually "it" is "they." *Convoluta roscoffensis* are flatworms within whose translucent bodies live grass-green algae. Annoyances to bathers,

Pan, a god of ancient Greece

they long baffled biologists. The flatworms and algae have merged into a composite creature. They lie in dense green masses on the shore, and instead of eating, make their own food from sunlight and air. They resemble plants until bothered by pounding surf or a predator, at which point they burrow for cover. The algae not only live inside the tissues of the flatworm and produce food for it but also recycle the worm's waste products, such as uric acid, into additional food. Due to this symbiotic relationship, adult worms do not have to eat and their mouths remain permanently closed.

Symbiosis—the living together in intimate association of different kinds of organisms—is more than an occasional oddity. It is a basic mechanism of evolutionary change. Some plants and animals would long ago have become extinct were it not for the help of their partners: blind shrimps are led around by sighted fish, flowering plants need to be pollinated by specific insects, cows and other ruminants cannot digest grasses without the aid of gut bacteria. Humans also need live bacteria in their intestines. We have trillions of animal cells—and ten times as many bacterial cells.

Although many plant and animal symbionts are known, symbiosis and its fundamental role in evolution really become conspicuous in the microcosm. Perhaps a hundred million years after mitochondria had become established, a new type of organism joined them in the cytoplasm of certain cells. But the genesis of the union was not through infection but ingestion.

Like Jonah swallowed by the whale, the forebears of the photosynthetic parts of nucleated cells were engulfed by larger bacteria but, far from being destroyed, found shelter within, resisted being digested, and kept their valuable light-trapping pigments alive. Today, locked inside every plant, these organelles, or plastids, make food from water and sunlight. Chloroplasts are green plastids and are even larger and more like bacteria than are the mitochondria. Plants turn toward sunlight because without it the plastids within would die.

Plastids provide the biosphere with food and oxygen. From a planetary point of view, the major role of mammals may be as fertilizers of plants and carriers of mitochondria. But if all mammals were to die in one instant, insects, birds, and other organisms would carry mitochondria and fertilize plants. If plants with their plastids were to suddenly disappear, the output of food on the planet would be so severely hampered that all mammals would certainly die.

A cell that didn't exist before would soon become indispensable to future generations. The new evolving cell now had mitochondria for oxygen metabolism and plastids to provide food. Both were the products of bacterial mergers. The question that remains is whether the cell's ability to move—even within its own cell wall—is the product of yet another symbiotic merger.

Whence Come the Tails of Sperm?

We believe that cellular motion by nucleated cells may be the result of a symbiotic merger between still other kinds of bacteria: rapid, whiplashing bacteria called spirochetes. Close study of the tiny cell whips on many kinds of cells with nuclei shows an amazing uniformity. These filaments have traditionally been called flagella if they are long and few like sperm tails, or cilia if they are short and numerous like hairs. Since there is no basic difference between them, they are all called undulipodia *(from the verb "undulate").* Nearly all algae and ciliates—the earliest organisms with nucleated cells to have evolved—have them. We are currently exploring the idea that undulipodia come from spirochetes, among the tiniest, fastest, most mobile members of the microcosm.

Shaped like corkscrews or bits of fusillini pasta, spirochetes thrive everywhere, from garden soil to people's gums. Some use oxygen; others are poisoned by it. They tend to attach to things, living or not. They form a major part of the microbial community that lives inside the swollen intestines of termites. There they can be seen attached to and feeding at the surfaces of larger organisms.

When, 2,000 million years ago, an organism with spirochetes propelling it found more food and reproduced more often, natural selection would have favored the alliance. A certain modern amoeba, for example, that draws in its whiptail and gorges itself when food is plentiful, grows a tail when food is scarce in order to swim in search of a meal. The advent of spirochete alliances would have altered the microcosm, leading to the first animal cells—a sort of symbiotic ménage à trois formed of *Thermoplasma*, mitochondria, and spirochetes. Plant cells may also be multispecies assemblies, composed of these plus plastids.

Proving the spirochete connection is difficult. As bacteria merge, promiscuous genes ultimately blend, and it becomes very difficult to sort out the original partners. The integrity of individual partners is sacrificed to the formation of a new cell. As David C. Smith puts it, what remains after the living merger, after billions of years of life within a supporting living habitat, is only the smile of Lewis Carroll's Cheshire cat: "the organism progressively loses pieces of itself, slowly blending into the general background, its former existence betrayed by some relic."

New techniques of molecular genetics confirm that parts of organisms dwindle within the life support system of other living cells. Bacteria can donate and receive varying numbers of genes, not only from each other but also from viruses and cells with nuclei. The free transfer of parts and pieces of living things from one area of a cell to another may explain how symbiotic organisms became streamlined into mere semblances of their former selves. The malleability of microbial life is exploited by genetic engineers who identify proteins they want to produce in large quantity, such as human insulin, and put the genes from them inside bacteria capable of rapid and prodigious reproduction. Not to belittle the human effort, it is worth pointing out that bacteria have been using "genetic engineering" techniques—transferring genes among themselves for their own purposes—for billions of years.

In the traditional view of a cutthroat Darwinian world, merged life forms have always seemed a bit odd, aberrations from the law of the jungle that the poet Tennyson characterized as "red in tooth and claw." Yet it now seems plants and animals never would have evolved at all were it not for attacks and defenses followed by symbiosis and reciprocity. Uneasy alliances are at the core of our very many different beings. Individuality, independence—these are illusions. We live on a flowing pointillist landscape where each dot of paint is also alive. Earth itself is a living habitat, a merger of organisms that have come together, forming new emergent organisms, entirely new kinds of "individuals" such as green hydras and luminous fish. Without a life-support system none of us can survive. It is in this light that we are beginning to see the biosphere not only as a continual struggle favoring the most vicious organisms but also as an endless dance of diversifying life forms, where partners triumph.

Five Dreamings (Michael Nelson Jakamarra, assisted by Marjorie Napaljarri; Central Australia, 1984). "We live on a flowing pointillist landscape where each dot of paint is also alive."—Lynn Margulis and Dorion Sagan

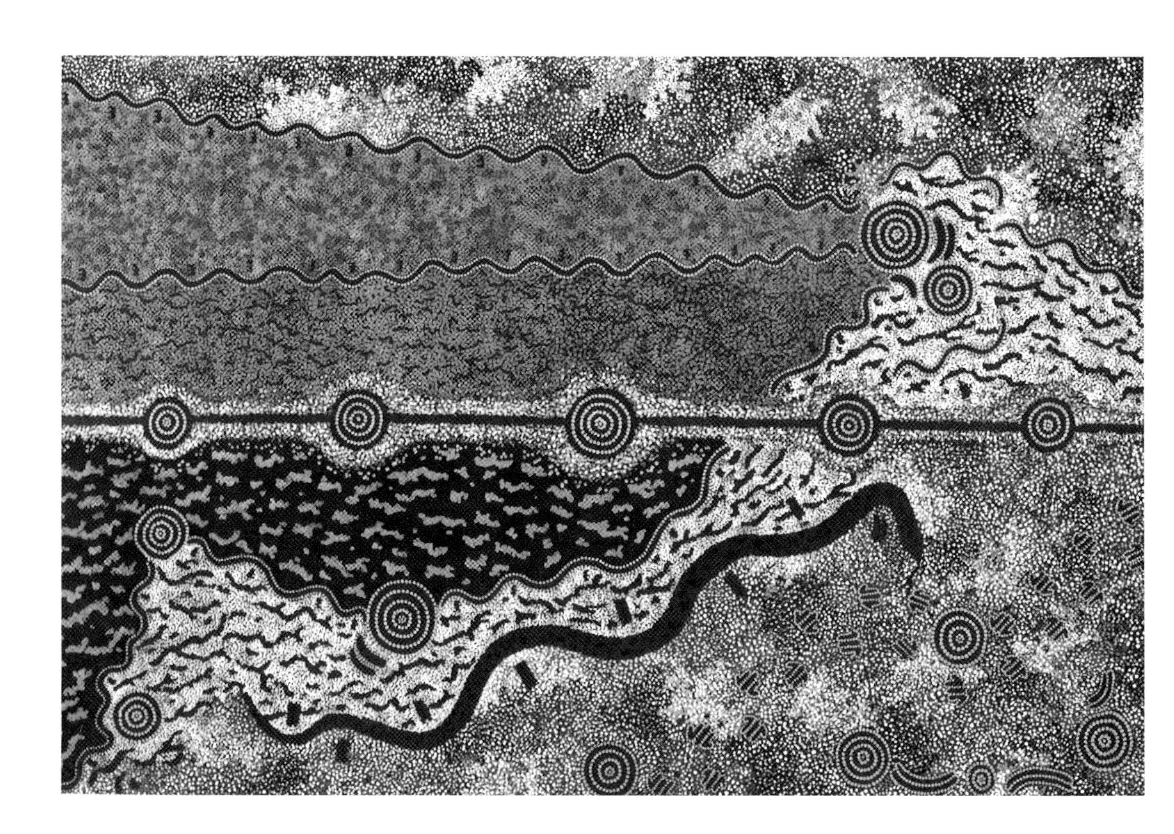

6 Blurred Bounds of Individuality

"Uneasy alliances are at the core of our very many different beings. Individuality, independence—these are illusions." Thus spoke Lynn Margulis and Dorion Sagan in the previous essay. While Margulis and Sagan call attention to the fuzzy boundaries between symbiotic species and between our organelles and ourselves, Julian Huxley here explores individuality within species. In his 1912 book, The Individual in the Animal Kingdom *(Cambridge University Press), Huxley draws upon examples ranging from colonial corals and clones of aphids to the identical quadruplets commonly born of armadillos.*

Julian Huxley was the grandson of Thomas Henry Huxley. The elder Huxley was a colleague of Charles Darwin and the most vociferous supporter during the late nineteenth century of Darwin's theory of evolution by natural selection. Two generations later, Julian Huxley was among a half dozen biologists who together reconciled the facts of paleontology and genetics with the theory of natural selection. Julian Huxley's name for this great effort has stuck; Darwinism today is sometimes called "the modern synthesis."

In The Individual in the Animal Kingdom *Julian Huxley surveys a dizzying number of cases in which delineation of individuals proves quite troublesome. As you will see, Huxley conducted his investigation not simply to catalog nature's curiosities but also to infuse a bit of science into traditional philosophical discourse. His essay here is by far the oldest of the works that appear in this anthology and his writing style is occasionally archaic, but you will see that his desire to bridge the intellectual realms is by no means out of date. Huxley concluded his preface with these words: "I will only hope that this little book may help, however slightly, to decrease still further the gap (to-day happily lessening) between science, philosophy, and the ideas and interests of everyday life."*

What Is an Individual?

Julian Huxley

"Accidents cannot happen to me." So says Nietzsche's Zarathustra, and in the saying proclaims to the world the perfection of his individuality. It might be thought that such a being was far outside the purview of the Zoologist, that he himself belonged to imagination and his individuality to the most speculative philosophy, and that both he and it should be left where they belong, where they could not contaminate the "pure objective truth of science."

That I think is an error: for the idea of individuality is dealt with of necessity both by Science and by Philosophy, and in such a difficult subject it would be mistaken to reject any sources of help. Not only that, but animal individuality with the advent of consciousness, though still remaining a lawful subject of the Zoologist, becomes naturalized in the proper realms of the Psychologist and the Philosopher and transfers thither the major portion of its business.

More, even were the Zoologist to confine himself to a description of non-conscious organic individuals and the deductions he drew from them, he would often find himself without a reasoned criterion of Individuality or a true idea of what he means by "higher" or "lower" individualities. It is only when the Biologist and the Philosopher join hands that they can begin to see the subject in its entirety.

When a glance is thrown over the various forms of animal life to which the name of Individual is naturally conceded, it is seen that in spite of many side-ventures, they can be arranged in a single main series in which certain characters are manifested more clearly and more thoroughly at the top than at the bottom. One of these characters is independence of the outer world and all its influences—in other words, immunity from accidents. By independence is not meant that independence of the recluse or the ascetic, but that other independence belonging to the great man of action and the inventor. These are not independent in the most literal sense—they do not "do without," they are not proud of existing on the barest minimum; the ultimate logical end of that kind of independence is atrophy, both mental and physical. Their other, higher independence involves this much of dependence, that they employ the things of the external world as material with which to work.

For the making of bricks, you are dependent upon straw: but you attain a higher independence by making bricks and being dependent on straw than by being independent of straw and lacking bricks. They gain their independence by using the outer world for their own ends, harnessing some of its forces to strive with and overcome the rest. At the least they can resist the adverse current, displaying a purpose of their own which is not whirled away by every wind of fate. "Accidents cannot happen to me"—so spake Zarathustra, and then added this reason: "Because all that could now happen to me would be my own."

We can deduce another attribute of individuality—its heterogeneity; from that very unity of the whole we can postulate diversity of its parts. This sounds paradoxical, but in reality it can be easily shown that nothing homogeneous can be an individual. In nonconscious organisms at least, difference of function always implies difference of structure, so

Procession of Monks
(Nantembō, 1924)

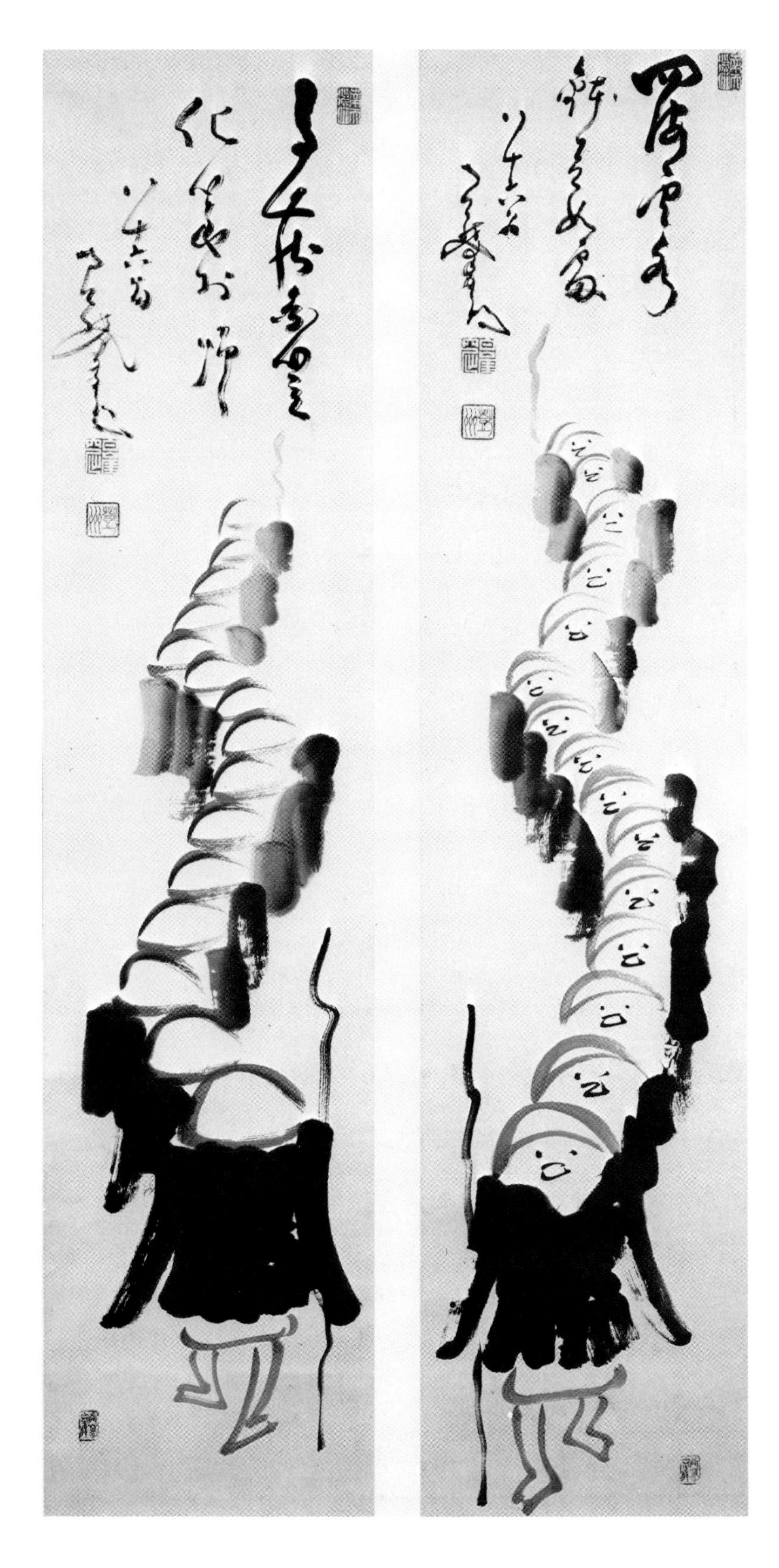

that the more independence—the more individuality—an individual is to possess depends very closely on the amount of heterogeneity of its parts. Look for instance at such an individual as a colony of Termites, its defence delegated to one caste, its nutrition to another, its reproduction to another; the various castes are specially adapted in their structure for their various functions. It is obvious at once that the queen with her vast swollen abdomen full of eggs is a much more effective reproducer than if she had retained any of the structure and mobility necessary to defend or look after herself. The soldiers again could not have been such powerful defenders of the colony if they were to have kept any of the delicacy of mandible required by the workers, the craftsmen.

Persistence

What has been said so far presupposes some degree of continuance in the individual; a survey of the various kinds of organic individuals shows this continuance to be common to them all, and that too in no limited measure, but as one of the fundamentals of their existence. This view of the individual, as a whole whose diverse parts all work together in such a way as to ensure the whole's continuance, or, as the evolutionist would say, whose structure and working have "survival-value," cannot stand without some qualification. There is death to be reckoned with; the survival is only temporary.

Under cover of the one word Death lie sheltered two separate notions—death of the substance, when the living protoplasm ceases to exist as such, and death of the individuality informing the substance. In man, both are inseparably connected; in many lower animals they are not. To take the simplest example: most Protozoa, such as Amoeba or Paramaecium, definite individuals both, feed and continually grow, and when they are grown to a certain maximum size, divide into two halves, each of which reorganizes itself into an individual resembling its "parent." Not a jot of substance has been lost: but one individuality has disappeared and two new ones are there in its place.

Break off a Begonia leaf and chop it into little bits; each bit reveals its latent power, sending roots downward, shoots upwards, and at the last becoming a self-sufficient whole. Through this regulatory power, Life has been able to save herself a tossing from her dilemma, escaping, like a Minoan acrobat, between the very horns: through it she has the possibility of reproduction.

The essence of reproduction is that one individual should create a new individual out of itself. The parent may persist, as in man, after the offspring has come into the world, or, as in Protozoa, may annihilate itself in the very act; that does not matter. What matters is that in every species there exists a succession of individuals in time, each one built up and working on a common plan.

Our first definition of the individual based on the idea of continuance can now be amended. We must not say that the individual is a whole whose parts work together in such a way as to ensure that this whole, and its working, shall persist; the individual only persists for a limited time. In spite of this, something does indefinitely continue, though it is but the kind, the species, and not the single individual itself.

These qualifications, universally applicable though they are to all individuals that we know on this earth, are still mere qualifications, not essential to the pure idea of individuality: the perfect individual would be eternal, subduer of time as well as of space. Since, through practical difficulties, Life has not been able to reach this perfection, she has had to content herself with the next best, continuance of the kind of individual instead of the individual itself.

The Species as Individual

The existence of a species or race, a procession of similar individuals each descended from a previous one, as well as of what we usually call individuals, the separate beings that at any one moment represent the species, leads of necessity to the separation of two distinct kinds of individuality; one belonging to the race and one to the persons that constitute the race. *(What Huxley states here so matter-of-factly is today a hot debate among philosophers of biology. In 1974 Michael Ghiselin published a paper that challenged the conventional wisdom. Ghiselin, like Huxley, argued that a species is a bona fide kind of individual. The debate that sprung from Ghiselin's paper is more than just semantic; it concerns fundamental notions of evolutionary biology.)*

Take as an example *Distomum hepaticum*, the Liver Fluke. The eggs of this unpleasant creature, which gives sheep the disease known as liver-rot, are passed out of the host and hatch out into minute embryos that swim about in the film of moisture on the meadow-plants. They cannot develop further unless they fall in with a particular sort of snail: if so, they burrow into its liver, and grow up, not into a new fluke, but into an irregular sort of bladder, the sporocyst; this, from its inner wall, produces a number of new embryos which grow and burst out of their parent as the so-called rediae—individuals differing both from the fluke or the sporocyst. These in turn give rise to a number of little tailed creatures, the cercariae, which migrate out of the snail, pass into a resting stage on blades of grass, and there passively await a browsing sheep. If one by good chance devours them, they hatch out, bore their way into the liver, and grow up again into flukes.

Now each of these three forms that thus cyclically recur is obviously an individual in the sense defined by us: they are wholes with diverse parts, whose working tends to their own continuance, even though this continuance is limited. But besides this there is the cycle itself to be

Sky and Water I (M. C. Escher, 1938)

reckoned with: it too is a definite something, a whole, it too is composed of diverse parts, sporocyst, redia, fluke, it too works in such a way that it continues (and continues indefinitely). What right have we to deny it any individuality as real as those possessed by any of its parts? True those parts are separated in space; but the ant-colony shows that this is no bar to individuality.

The real point is this: the existence of the sporocyst and the redia is of no direct advantage to the individual fluke: it would grow and lay eggs just as happily if the host-snails, and with them all the sporocysts and rediae, present and to come, were exterminated. It is however of advantage to something, and that something can only be the race of liver-flukes, the kind of protoplasm which by its difference from other kinds has earned a special name—*Distomum hepaticum.* That is an extreme case; the two kinds of individuality may often be inextricably interwoven. What is of advantage to one is usually of advantage to the other, so that, by an over-emphasis of the species-individuality of which we are the parts, it is often said that our bodies are only "cradles for our germ-cells."

It must here suffice to say that wherever a recurring cycle exists (and that is in every form of life) there must be a kind of individuality consisting of diverse but mutually helpful parts succeeding each other in time, as opposed to the kind of individuality whose parts are all co-existent. The first constitutes what I shall call species individuality, or individuality in time, while the other corresponds to our ordinary notions of individuality and, if a special term is needed, may be called simultaneous or spatial individuality.

The continuance of the working of a species as we have defined it would preclude change; but change and the idea of evolution are at the base of all modern thought in science and philosophy alike. To-day few would be found to deny that all the battalions of living organisms are descended from one primeval type. That is the logical outcome of the doctrine of Evolution. Evolution is a word glibly used, but often without thought of its full meaning. If Evolution has taken place, then species are no more constant or permanent than individuals. As individual emerges from individual along the line of species, so does species emerge from species along the line of life, and every animal and plant, in spite of its separateness and individuality, is only a part of the single, continuous, advancing flow of protoplasm that is invading and subduing the passive but stubborn stuff of the inorganic.

Consciousness

In the actual duration of his life, the individual ranges from the bacterium's hour to the big tree's five thousand years. *(Recent discoveries of field biologists indicate that the sequoias and bristlecone pines of California may not be the oldest individual plants. A clone of blueberries blanketing an Appalachian mountaintop and a clone of aspens on fire-prone slopes in the Rockies may have rootstocks ten or fifteen thousand years old.)* Man in this again stands on the pinnacle of individuality—not in mere length of days, but in having found a means to perpetuate part of himself in spite of death.

By speech first, but far more by writing, and more again by printing, man has been able to put something of himself beyond death. In tradition and in books an integral part of the individual persists, and a part which still works and is active, for it can influence the minds and actions of other individuals in different places and at different times: a row of black marks on a page can move a man to tears, though the bones of him that wrote it are long ago crumbled to dust. In truth, the whole of the progress of civilization is based on this power. Once more the upward progress of terrestrial life towards individuality has found apparently insurmountable obstacles, gross material difficulties before it, but once more through consciousness it finds wings, and, laughing at matter, flies over lightly where it could not climb.

To such an individuality, one that can thus transcend the limits of its substance, the name of Personality is commonly given. Man alone possesses true personality, though there is as it were an aspiration towards it visible among the higher vertebrates, stirring their placid automatism with airs of consciousness. *(Huxley's portrayal of a kind of upward striving or progress in evolution is decidedly unfashionable among evolution theorists today. Stephen Jay Gould in particular has devoted great effort toward expunging this popular notion of inevitably progressive evolution.)*

In man, personality is usually defined with reference to self-consciousness rather than to individuality; but the power of reflection and self-knowledge is linked up, in our one type of personality at least, with the new flight of individuality—conscious memory seems necessarily to imply a vast increase of independence, so that it is all one whether we define the possessor of a personality as a self-conscious individual, or as an individual whose individuality is more extensive both in space and time than the material substance of its body.

Personality, as we know it, is free compared with the individuality of the lower animals; but it is still weighted with the body. There may be personalities which have not merely transcended substance, but are rid of it altogether: in all ages the theologian and the mystic have told of such "disembodied spirits," postulated by the one, felt by the other, and now the psychical investigator with his automatic writing and his cross-correspondences is seeking to give us rigorous demonstration of them. If such exist, they crown Life's progress; she has started as mere substance without individuality, has next gained an individuality co-extensive with her substance, then an individuality still tied to substance but transcending it in all directions, and finally become an individuality without substance, free and untrammeled. *(Julian Huxley was a strong proponent of Teilhard de Chardin's vision of evolution as progress culminating in a nonmaterial "noosphere." In 1958 Huxley—by then Sir Julian Huxley—wrote the introduction to the English-language edition of Teilhard's* The Phenomenon of Man.*)*

That for the present must be mere speculation. The Zoologist has strayed: he must return to his muttons and his amoebae, and in the next chapter will begin to consider more closely the actual facts of animal individuality and their probable explanation.

Animals as Colonies of Cells

There are communities, such as those of bees and ants, where, though no continuity of substance exists between the members, yet all work for the whole and not for themselves, and each is doomed to death if separated from the society of the rest. There are colonies, such as those of corals or of Hydroid polyps, where a number of animals, each of which by itself would unhesitatingly be called an individual, are found to be organically connected, so that the living substance of one is continuous with that of all the rest. Sometimes these apparent individuals differ among themselves and their energies are directed not to their own particular needs, but to the good of the colony as a whole. Which is the individual now?

Histology then takes up the tale, and shows that the majority of animals, including man, our primal type of individuality, are built up of a number of units, the so-called cells. Some of these have considerable independence, and it soon is forced upon us that they stand in much the same general relation to the whole man as do the individuals of a colony of coral polyps, or better of Siphonophora *(jellyfish)* to the whole colony. This conclusion becomes strengthened when we find that there exist a great number of free-living animals, the Protozoa, including all the simplest forms known, which correspond in all essentials, save their separate and independent existence, with the units building up the body of man. Both, in fact, are cells, but while the one seems to have an obvious individuality, what are we to say of the other?

From One to Two: Budding and Fission

So far we have treated the problem statically, as it were: when we come to view it dynamically, tracing the movement of life along its course, the difficulties do but increase. Take, to begin with, a simple colony of hydroid polyps, and ask how does this multiplicity of connected animals arise? Observation shows the whole stock to be formed, by a process of budding, from one original individual. A little lump or knob is seen at one place, which, growing rapidly, bit by bit assumes the appearance of the individual whence it has sprung.

Ernst Haeckel drew these sea anemones in the late nineteenth century. These marine invertebrates, like their hydroid relatives, reproduce both sexually and by budding.

By this means, the first individual produces a second out of itself. Its own individuality is not lost in the process. The relation of the individuals in a colony to each other is thus rendered still more obscure owing to the fact of one being produced out of another. What was at first nothing but a part grows up into a new whole.

Budding, though perhaps most striking when it leads to the formation of a colony, is by no means restricted to colonial forms. Often, as in Hydra, the process is completed, and the bud set free to lead an independent life. Here one individual has produced a second out of its own substance; the two resemble each other not less closely than two individuals bred from the egg, and yet the first has lost not a jot of its own individuality in thus creating itself anew in the second.

This fresh creation of new forms from the substance of the old is what we usually term Reproduction. Budding is but one of its many methods, and we must look at some others before we can see its full bearing upon our subject. First we will take fission, or division into two halves, a method which occurs in several groups of higher animals, though less commonly than budding. Rarely, as among the stony corals, are colonies produced through its means; usually the two halves part company and each becomes as perfect an individual as its parent. It is, however, in this relation of parent to offspring that division is at variance with budding. Instead of one individual producing another, here the founder of the race ceases to exist, losing his own individuality in the production of two fresh ones. Division is thus even more important for the present purpose than budding; we have the strange paradox that though each individual hands on the whole of its substance intact to its successors, yet with this perfect continuity of substance there co-exists perfect discontinuity of individualities.

Sexual Reproduction

There remains the third chief mode of reproduction. In considering the hydroid colony, we found that all its members took their origin by budding from one single founder. This founder, though identical in organization with the rest, has yet not had the same origin as they. Tracing its life backward towards its source, we find it first of smaller size; then comes the stage when its organs are developed one by one, much as in the bud; before this it exists in a form through which the budded individuals never pass—as a small drawn-out ovoid, actively swimming instead of fixed to the ground; before this again it is seen as a round motionless body, built up like a mulberry out of rounded parts, and finally, its "fount of life" is revealed in a spherical inert mass, single and undivided—the fertilized egg.

This fertilized egg is neither more nor less than a cell—specialized, as one would expect, for the discharge of its own particular duties, but still a cell. Here is a further strengthening of our view of the higher animal

or metazoan as a colony of units each comparable with a protozoan. When the method is traced by which the plurality of cells in the adult arises from the single cell of the egg—the method, that is, of cell reproduction—it is found to be identical with one of the ways of reproduction in metazoan individuals, that of fission; the single founder of the cell-community, the egg, divides the whole of its substance into two halves, each of which is a new cell. This is repeated again and again, and the whole army of cells in the full-sized hydroid are direct descendants of that single founder-cell. But the hydroid is itself a founder; and the new "individuals" which it buds out depend for their growth upon this same process of cell-division continually repeated.

The paradox is growing yet. Each hydroid seems in its way a whole; yet it is as well a mere part of a single greater whole, the colony, and, besides this, itself composed of units each of which again is in some sort a whole: and each whole has some claim to the name of Individual.

One gap still yawns: what was the origin of the single cell that gave birth to the whole adult organism? In this particular case, it was a fertilized ovum: by which is implied that the single cell has arisen from the total fusion, body and soul, or rather cytoplasm and nucleus, of two other cells; these are technically known as the gametes, and their product, the fertilized ovum, as the zygote. These two cells have come from two separate individual persons (one male and one female) and their cell-ancestors have been firmly built into the fabric of these individuals' bodies. *(Many marine invertebrates, and even some fishes, are neither male nor female; they are hermaphrodites, producing both eggs and sperm.)* This merging of two cells and their two individualities in one (the exact reverse of fission) is the essential sexual act.

Individuality and Identical Twins

One of the most widespread definitions considers the individual as "the total product of a single impregnated ovum," that is to say as the sum of the forms which appear between one sexual act and the next. This would make all the polyps in a colony of hydroids, all the separate polyps budded off by a fresh-water hydra, all the summer generations of the aphis, together constitute but a single individual. Of the various facts which make the hypothesis untenable, the chief are concerned with the artificial or accidental production of two or more co-existent organisms from a single ovum.

In most animals each single fertilized egg gives rise to a single embryo and this to a single adult organism; but in some, where this is the normal rule, more than one embryo may be accidentally or artificially formed from one egg, and in others this multiplicity is the usual course of events, even though most of their relations may grow up in the ordinary humdrum way—"one egg, one adult." Aberrations may occur even in man; there can be very little doubt that identical twins arise from the two cells produced by the first division of a single fertilized ovum, which have accidentally been torn apart instead of staying united. A very interesting variation on this is seen in the nine-banded armadillo which regularly produces "identical quadruplets." Most mammals give birth to several young at one time, but usually each grows up from a separate and separately fertilized ovum and each is enclosed in its own set of embryonic membranes. The armadillo's brood, however, like the identical twins in man, has only a single chorionic membrane, and the four resemble each other minutely. Always of the same sex, their measurements are identical; even the number of plates in their armour is constant to less than one per cent., though the range of variation from brood to brood may be five percent and more.

Then comes Experiment and confirms our conclusions of observation. The egg when it develops outside the body of its parent (the rule with most of the lower animals) is at the mercy of the experimenter. After it has divided into two halves, these two blastomeres (as the cells produced by the subdivision of the egg are called) can be separated either mechanically or by chemical means. In the majority of animals where this is possible, the half-blastomere, that identical mass of substance which without man's intervention would have formed half the body of the adult, develops, owing to the mere accident of separation from its sister, into a whole body. Even with such a highly organized creature as the newt this has been accomplished.

It is difficult to consider the two experimentally produced newts as constituting a single individual; the four armadilloes with their one individuality raise more than a doubt; and with the occasional and accidental production of true twins in man comes finality. If anything is an individual on this earth, that surely is man; and yet we are asked to believe that though the most of us are true individuals, yet here and there some man who lives and moves and has his being like the rest is none, that he must make shift to share an individuality with another man simply because the couple happen to be descended from one fertilized egg instead of two. In himself a twin is like any other man; to say that one is an individual while the other is not, takes all the meaning from the word.

From One to Two: Regeneration

To add the final straw, regeneration comes. Regeneration is usually looked on as something strange, almost abnormal, owing to its not occuring in man or his animal familars. In reality it is much rather an original property of life, which for special reasons has dropped out of the human scheme of things.

As we descend the vertebrate scale, it is not until we reach the lower Amphibia, such as the newt and salamander, that regeneration becomes at all marked. Even here, it is present in restricted form, and is confined to the restoring of lost organs. A leg, that is to say, or a tail, even an eye or a jaw may be replaced, but the central systems and main lines of organization must be left intact. There must remain a certain central residue of the individual if it is to complete itself.

This in itself points to a vaguer, more fluid notion of individuality than can ever be got from contemplation of man alone, but what are we to say of such things as happen in many of the lower animals? Take first one form of regeneration seen in Clavellina, one of those poor relations of Vertebrates, the Ascidians *(commonly called tunicates or sea squirts).* Cut Clavellina across in the middle, and (in certain defined conditions) a bud will sprout from the front end of the hinder half, and another from the hinder end of the front half. As in the growth of the hydroid colony, the old organization is kept entire and whole, the new organization is built up in the bud; here, however, it is not one whole individual giving rise to another, but one half giving rise to just that dissimilar half which is its complement. There is an old school-boy question about a cricket bat: suppose the handle of a bat broke, and a new one was put on to the old blade. Suppose then that the blade broke and was in its turn replaced; would the bat still be the same bat? That is a hard question, but Clavellina asks a harder still.

Man Is Not the Measure

By now, all faith in man as a guide to individuality must have been shattered. In man, individuality presents itself as something definite and separate from all others, something which animates a particular mass of matter and is inflexibly associated with it, appearing when it appears and vanishing when it dies. That idea of individuality is not universally applicable.

In perplexing procession before us there have appeared individualities inhabiting single cells, others inhabiting single cells at the start, many cells in later life; individualities whose fleshly mansions are continuous one with another, no boundaries between; individualities that appear and disappear along an undying stream of substance, the substance moulding itself to each as the water of a stream is moulded in turn to each hollow of its bed; within one individuality others infinite in number, lying hid under the magic cloak of potentiality, but each ready to spring out as if from nowhere should occasion offer. Nothing remains but to abandon preconceived ideas. We must seek to interpret human individuality not as the one true pattern to which all others must conform, but as something with a history and intelligible only through that history.

Individuality in "the whole organic world"

Green plants can build up protoplasm from water, carbon dioxide, and mineral salts; the protoplasm thus formed is the ultimate source of all nourishment to the rest of life. Animals either eat green plants or else eat other animals that eat green plants; many bacteria feed on the dead tissues of animals and plants, bringing about as a result of their activity the phenomenon known as decay; and fungi live to a great extent on the substances produced during decay. Meanwhile, however, the waste products of the current of metabolism and the final products of decay, which come eventually to be degraded to simple stable substances such as water, carbon dioxide, ammonia, and nitrates, get diffused in the soil, and form the basis once more of the green plant's activity.

In a sense, therefore, the whole organic world constitutes a single great individual, vague and badly coordinated it is true, but none the less a continuing whole with inter-dependent parts. Within this biggest system, nature has been persistent in her efforts to create other "naturally-isolated systems," other individualities. Out of every little accidental company she tends to make an inter-related whole whose parts are largely dependent on each other, and only slightly on other wholes or their parts.

Social Organizations as Individuals?

A man can very well be at one time a member of a family, a race, a club, a nation, a literary society, a church, and an empire. "Yes, but surely these are not individuals,"—I seem to hear my readers' universal murmur. That their individuality is no mere phantasm I think we must own when we find men like Dicey and Maitland admitting that the cold eye of the law, for centuries resolutely turned away, is at last being forced to see and to recognize the real existence, as single beings that are neither aggregates nor trusts, of Corporate Personalities. *(In 1886 the U.S. Supreme Court ruled that a business corporation is indeed a "person" and thereby subject to the same Constitutional protections afforded other persons. In the decades following this historic ruling, the growth of corporations boomed. Corporations that allowed their shareholders to trade or sell stock freely, and which protected those shareholders with the assurance of "limited liability" against corporate misdeeds that might bring criminal or civil penalties and perhaps bankruptcy, found it easier to attract the capital necessary for growth. Legal changes enabling business associations to transform themselves from mere aggregations of human individuals into a kind of superindividuality thus led to a boom in economies so organized.)*

This being so, it yet remains that the state or society at large is still a very low type of individual; the wastage and friction of its working are only too prominently before our eyes. With the examples of what life has accomplished in producing our own bodies, we can never despair. But we must not be too far tempted by biological analogies; the main problem is the same, but the details all are new. The individuals to be fused into a higher whole are separate organisms with consciousness, reasoning minds—personalities; and the solution will never be found in the almost total subordination of the parts to the whole, as of the cells in our own bodies or the sweated laborer in our present societies, but in a harmony and a prevention of waste, which will both heighten the individuality of the whole and give the fullest scope to the personalities of all its members.

Individuality and the Units of Selection

Although Julian Huxley framed his ideas long ago, the issue of individuality is still a hot topic in biology. Population Biology and Evolution of Clonal Organisms, *edited by Jeremy B. C. Jackson, Leo W. Buss, and Robert E. Cook (copyright 1985 by Yale University Press), begins with this statement (reprinted by permission of Yale University Press).*

Jackson, Buss, and Cook

Clonal growth—the formation of more than one individual of identical genetic composition—is a fundamental ecological adaptation that has far-reaching consequences for the population biology, morphology, development and evolution of such organisms. Ever since Darwin, the development of theory in ecology and evolution has been implicitly constructed for fruit flies, birds, and people, unitary organisms whose populations comprise readily distinguishable, sexually derived individuals of approximately determinate adult size and life span. Grasses, vines, sponges, corals, and other clonal organisms, which commonly dominate much of the land and sea and do not commonly display such characteristics, have been largely ignored by theorists.

Extensive recent interest in the ecology, morphology, and genetics of clonal organisms has clearly demonstrated the inadequacy of established concepts of population biology for study of clonal species. This problem has not only held back our understanding of the dynamics and evolution of more than half the biosphere; it has also muddled understanding of other substantive but unresolved issues regarding all organisms, such as the evolution of sex and senescence.

In The Evolution of Individuality *(copyright 1987 by Princeton University Press), Leo W. Buss attempts to broaden "the modern synthesis" forged in Julian Huxley's era. While these earlier biologists did find a way to unite the facts of paleontology and genetics with the theory of natural selection, Buss criticizes them for leaving out one key area: ontogeny (sometimes called embryology), which pertains to the growth and development of organisms in between those spurts of sexual reproduction that punctuate the generations. Because of this omission, Buss maintains, Huxley and his generation of biologists failed to formulate a question fundamental to a full understanding of evolution.*

Leo W. Buss

The synthetic theory has succeeded because evolution has acted in all cellular-differentiating organisms in a manner such that individuality is approximated, albeit to varying degrees. Organisms appear as individual

A single clone of aspen trees in the Rocky Mountains of Colorado

entities; evolution has yielded unambiguously discrete units. The Modern Synthesis has stood as an enduring intellectual edifice for nearly a half-century because individuality has evolved.

Why, though, has individuality evolved? Lacking in the Modern Synthesis, in evolutionary theory altogether, is an approach to this problem. Why, for example, does a mouse sequester its germ line and a hydroid not? Why do polychaetes have a variable number of segments and leeches a constant number? The Modern Synthesis has not generated a theory of ontogeny. Nor can it be expected to generate one: a theory which assumes individuality as a basal assumption cannot be expected to explain how individuality evolved.

The history of ideas is paved by the constraints of language. How we communicate frames what we can communicate. How we communicate frames what we see as trivially true and what we see as inherently paradoxical. How we communicate frames, in subtle and insidious ways, our very concepts of knowledge and hence what we claim to know. As scientists we are typically free to function successfully in blissful ignorance of such issues. We share a common language, acquired through a long apprenticeship and jealously safeguarded by the social strictures of peer review. It is rare indeed that a formative concept falls on such hard times that a new language must be developed to take the place of verbal conventions whose time has passed. But such a time is upon us.

Few would debate that the Modern Synthesis, built on the explicit assumption that all evolutionary change is attributable to variation among individuals, is shortly due for a considerable theoretical expansion in scope. Likewise, few would doubt that this expansion will, in some pivotal way, focus on what has been called the "units of selection problem." What remains in doubt is the language that we will use.

The units of selection that evolutionists of the twentieth century most frequently debate—individuals, populations, and species—are only the latest in a long history. That the century following Darwin should have directed its attention to these units and primarily to selection at the level of the individual is hardly surprising. The theory of evolution was originally framed largely in terms of individuals because individuals were the simplest entities for a nineteenth-century naturalist to work with and, presumably, to think about. To confuse this matter of expediency with sufficient evolutionary theory, however, would be a serious error. In previous eras, self-replicating molecules, complexes of such molecules, and cells were sequentially the units of selection acted upon by the external environment, and a theory of evolution must accommodate selection on these units and the transitions between these units.

Buss thus portrays the units of selection as a question central to current debate in evolution theory, and he goes on to propose rather startling approaches to the problem, which are, however, too technical to set forth here. In chapter 14 you will encounter Richard Dawkins' ideas on this topic. Fortunately, Dawkins has geared his seminal work on the selfish gene theory for a popular as well as a scientific audience, and you will see that his ideas (and the units-of-selection debate in general) do have profound implications for how we view the world, indeed, the very meaning of life.

The following essay first explores a view of nature that Buss advocates as essential for tackling the units-of-selection problem and for piercing the remaining mysteries in evolution theory. Buss claims that a "hierarchical perspective" is key. Arthur Koestler's essay is the clearest and most playful introduction to hierarchy theory that I have seen. With his usual flair, Koestler not only explains the science but draws dramatic connections between points of science and problems of society.

III *A Systems View of Life*

Thus, the task is not so much to see what no one yet has seen, but to think what nobody yet has thought about that which everybody sees.

—Schopenhauer

No man is an island—he is a holon. A Janus-faced entity who, looking inward, sees himself as a self-contained unique whole, looking outward as a dependent part.

—Arthur Koestler

7 *Holons and Hierarchy Theory*

Trespasser in an Age of Specialists

The Annual Obituary 1983

"I think that there are seasons in life," Arthur Koestler once said, "that there are moments in which you have to do something new, turn toward other pastures." Turn toward them he did: Arthur Koestler was a novelist, essayist, journalist, philosopher of science, historian, and political activist. To his peers, he was the writer who shattered the myths of the socialist revolution in Russia with his powerful novel, *Darkness at Noon*. To younger generations, intrigued by his theories of mind and parapsychology, Koestler was a writer who tried to integrate mysticism with science. To Koestler himself, he was a "trespasser in the age of specialists." What linked his work, fiction and nonfiction alike, was an unwavering search for a rational understanding of the human condition, of a world in which humanity seems hell-bent for self-destruction.

So begins the article on Arthur Koestler in The Annual Obituary 1983 *(Chicago and London: St. James Press, 1984). In late winter 1983, suffering from Parkinson's disease and leukemia, the 78-year-old Koestler took an overdose of barbituates. Koestler thus practiced what he preached. Some of his last writing had been for the British organization EXIT, a group dedicated to "the right to die with dignity." In a eulogy upon his death, one friend recalled Koestler's words that death is "merging into cosmic consciousness, . . . the flow of a river into the ocean. The river has been freed of the mud that clung to it and regained its transparency. It has become identified with the sea, spread over it, omnipresent in every drop, catching a spark of the sun."*

The obituary continues: As a very young man and ardent Zionist, Koestler went to Palestine, living in poverty in a small collective in what is now Israel. Then, in the early 1930s, Koestler became a secret member of the Communist Party, traveling to war-torn Spain (under the guise of a journalist) to investigate German and Italian aid for Franco. There he was imprisoned and nearly faced death by firing squad. Later abandoning the Communist Party, Koestler embarked on a twenty-year campaign to warn the world of the dangers of totalitarian governments. In the late 1950s he spearheaded a movement in his adopted country, Great Britain, to abolish the death penalty. Age neither cooled the passions nor slowed the writing. In all, Koestler wrote six novels, thirty books of nonfiction, and countless articles and essays for newspapers worldwide.

One of the central problems which preoccupied Koestler was the dilemma of means and ends—around which of these can the best political system be developed? Koestler's metaphor for this either/or problem was the Yogi and the Commissar. The Yogi represents the ideal of pure means, believing that social change can effectively come only from internal change. In the Yogi's worldview, the future will always remain an unknown quantity; external manipulation of society by a government in order to promote change leads only to chaos. On the other hand, the Commissar is ends-oriented. In his view, the future is deterministically knowable. Unlike the yogi, he is enmeshed in political activism, seeking to manipulate society toward his desired ends. This duality would reappear in various manifestations throughout Koestler's work and life. "I've always tried to reconcile the yogi and the commissar within me," he said. "It's not easy."

"To really understand politics, you must study not only history but psychology. . . . You can not discuss such matters without examining human evolution and the brain's neuro-physiology—in short, how man became what he is. So, my books from 1955 on have dealt with man's glory and predicament. Particularly in my trilogy (*The Sleepwalkers*, *The Act of Creation*, *The Ghost in the Machine*) I've tried to account for man's creativity and his genocidal madness, the two sides of the coin minted by the evolutionary process."

On the topic of creativity, here is my favorite passage from The Sleepwalkers *(published by Hutchinson Publishing Group Ltd. in 1959 and reprinted here by permission of the Peters Fraser & Dunlop Group Ltd.):*

Arthur Koestler

Most geniuses responsible for the major mutations in the history of thought seem to have certain features in common; on the one hand scepticism, often carried to the point of iconoclasm, in their attitude towards traditional ideas, axioms and dogmas, towards everything that is taken for granted; on the other hand, an open-mindedness that verges on naive credulity towards new concepts which seem to hold out some promise to their instinctive gropings. Out of this combination results that crucial capacity of perceiving a familiar object, situation, problem, or collection of data, in a sudden new light or new context: of seeing a branch not as part of a tree, but as a potential weapon or tool; of associating the fall of an apple not with its ripeness, but with the motion of the moon. The discoverer perceives relational patterns or functional analogies where nobody saw them before, as the poet perceives the image of a camel in a drifting cloud.

Every creative act—in science, art or religion—involves a regression to a more primitive level, a new innocence of perception liberated from the cataract of accepted belief.

Arthur Koestler did more than just interpret the scientific endeavor; he sought to advance it. His several trespasses into science were respected by some insiders and snubbed by others. Hierarchy theory, for example, is today embraced by some biologists, including Leo W. Buss, whose work appeared in the previous chapter. Koestler did not invent hierarchy theory; roots are apparent even in Julian Huxley's 1912 book, and in the writings of several of Huxley's contemporaries. But as you will see, Koestler did pull together existing ideas to form a coherent worldview. Since then biologists have drawn out of the theory some methodoligical insights that aid research. (See the bibliography.)

Passages below are drawn from Koestler's 1967 book, The Ghost in the Machine, *and from his 1978 book,* Janus: A Summing Up. *Both are published by Hutchinson Publishing Group Ltd. and reprinted by permission of the Peters Fraser & Dunlop Group Ltd. (U.S. rights for* Janus *are granted by Random House, Inc.)*

Parable of the Two Watchmakers

Arthur Koestler

Let me start with a parable. I owe it to Professor Herbert A. Simon, designer of logic computers and chess-playing machines, but I have taken the liberty of elaborating on it.

There were once two Swiss watchmakers named Bios and Mekhos, who made very fine and expensive watches. Their names may sound a little strange, but their fathers had a smattering of Greek and were fond of riddles. Although their watches were in equal demand, Bios prospered, while Mekhos just struggled along; in the end he had to close his shop and take a job as a mechanic with Bios. The people in the town argued for a long time over the reasons for this development and each had a different theory to offer, until the true explanation leaked out and proved to be both simple and surprising.

The watches they made consisted of about one thousand parts each, but the two rivals had used different methods to put them together. Mekhos had assembled his watches bit by bit—rather like making a mosaic floor out of small coloured stones. Thus each time when he was disturbed in his work and had to put down a partly assembled watch, it fell to pieces and he had to start again from scratch.

Bios, on the other hand, had designed a method of making watches by constructing, for a start, sub-assemblies of about ten components, each of which held together as an independent unit. Ten of these sub-assemblies could then be fitted together into a sub-system of a higher order; and ten of these sub-systems constituted the whole watch. This method proved to have two immense advantages.

In the first place, each time there was an interruption or a disturbance, and Bios had to put down, or even drop, the watch he was working on, it did not decompose into its elementary bits; instead of starting all over again, he merely had to reassemble that particular sub-assembly on which he was working at the time; so that at worst (if the disturbance came when he had nearly finished the sub-assembly in hand) he had to repeat nine assembling operations, and at best none at all. Now it is easy to show mathematically that if a watch consists of a thousand bits, and if some disturbance occurs at an average of once in every hundred assembling operations—then Mekhos will take four thousand times longer to assemble a watch than Bios. Instead of a single day, it will take him eleven years. And if for mechanical bits we substitute amino acids, protein molecules, organelles, and so on, the ratio between the time-scales becomes astronomical; some calculations indicate that the whole lifetime of the earth would be insufficient for producing even an amoeba—unless he becomes converted to Bios' method and proceeds hierarchically, from simple sub-assemblies to more complex ones.

A second advantage of Bios' method is of course that the finished product will be incomparably more resistant to damage, and much easier to maintain, regulate, and repair, than Mekhos' unstable mosaic of atomic bits. We do not know what forms of life have evolved on other planets in the universe, but we can safely assume that wherever there is life, it must be hierarchically organised.

The word "hierarchy" is of ecclesiastical origin and is often wrongly used to refer merely to order of rank—the rungs on a ladder, so to speak. I shall use it to refer not to a ladder but to the tree-like structure of a system, branching into subsystems. The concept of hierarchic order occupies a central place in this book, and lest the reader should think that I am riding a private hobby horse, let me reassure him that this concept has a long and respectable ancestry. So much so, that defenders of orthodoxy are inclined to dismiss it as old hat—and often in the same breath to deny its validity. Yet I hope to show as we go along that this old hat, handled with some affection, can produce lively rabbits.

Enter Janus

If we look at any form of social organisation with some degree of coherence and stability, from insect state to Pentagon, we shall find that it is hierarchically ordered. The same is true of the structure of living organisms and their ways of functioning—from instinctive behavior to the sophisticated skills of piano-playing and talking. And it is equally true of the processes of becoming—phylogeny *(evolution)*, ontogeny *(growth from embryo to adult)*, and the acquisition of knowledge. However, if the branching tree is to represent more than a superficial analogy, there must be certain principles or laws which apply to all levels of a given hierarchy, and to all the varied types of hierarchy just mentioned—in other words, which define the meaning of "hierarchic order."

The first universal characteristic of hierarchies is the relativity, and indeed ambiguity, of the terms "part" and "whole" when applied to any of the sub-assemblies. A "part," as we generally use the word, means something fragmentary and incomplete, which by itself would have no legitimate existence. On the other hand, a "whole" is considered as something complete in itself which needs no further explanation. But "wholes" and "parts" in this absolute sense just do not exist anywhere, either in the domain of living organisms or of social organisations. What we find are intermediary structures on a series of levels in an ascending order of complexity: sub-wholes which display, according to the way you look at them, some of the characteristics commonly attributed to wholes and some of the characteristics commonly attributed to parts.

The members of a hierarchy, like the Roman god Janus, all have two faces looking in opposite directions: the face turned towards the subordinate levels is that of a self-contained whole; the face turned upward towards the apex, that of a dependent part. One is the face of the master, the other the face of the servant. This "Janus effect" is a fundamental characteristic of sub-wholes in all types of hierarchies.

But there is no satisfactory word in our vocabulary to refer to these Janus-faced entities: to talk of sub-wholes (or sub-assemblies, substructures, sub-skills, sub-systems) is awkward and tedious. It seems preferable to coin a new term to designate these nodes on the hierarchic tree which behave partly as wholes or wholly as parts, according to the way you look at them. The term I would propose is "holon," from the Greek "holos" = whole, with the suffix "on" which, as in proton or neutron, suggests a particle or part.

"A man," wrote Ben Jonson, "coins not a new word without some peril; for if it happens to be received, the praise is but moderate; if refused, the scorn is assured." Yet I think holon is worth the risk, because it fills a genuine need. It also symbolises the missing link—or rather series of links—between the atomistic approach of the Behaviourist and the holistic approach of the Gestalt psychologist.

The Gestalt school has considerably enriched our knowledge of visual perception, and succeeded in softening up the rigid attitude of its opponents to some extent. But in spite of its lasting merits, "holism" as a general attitude to psychology turned out to be as one-sided as atomism was, because both treated "whole" and "part" as absolutes, both failed to take into account the hierarchic scaffolding of intermediate structures of subwholes. If we replace for a moment the image of the inverted tree by that of a pyramid, we can say that the Behaviourist never gets higher up than the bottom layer of stones, and the holist never gets down from the apex.

The two-term part-whole paradigm is deeply ingrained in our unconscious habits of thought. It will make a great difference to our mental outlook when we succeed in breaking away from it.

Holons in Social Systems

No man is an island—he is a holon. A Janus-faced entity who, looking inward, sees himself as a self-contained unique whole, looking outward as a dependent part. His self-assertive tendency is the dynamic manifestation of his unique wholeness, his autonomy and independence as a holon. On the level of the individual, a certain amount of self-assertiveness—ambition, initiative, competition—is indispensable in a dynamic society. At the same time, of course, he is dependent on, and must be integrated into, his tribe or social group. The integrative tendency expresses his dependence on the larger whole to which he belongs. The polarity of these two tendencies, or potentials, is one of the leitmotivs of the present theory. Empirically, it can be traced in all phenomena of life; theoretically, it is derived from the part-whole dichotomy inherent in the concept of the multi-layered hierarchy.

The manifestations of the two tendencies on different levels go by different names, but they are expressions of the same polarity running through the whole series. The self-assertive tendencies of the individual are known as "rugged individualism," competitiveness, etc.; when we come to larger holons we speak of "clannishness," "cliquishness," "class-consciousness," "esprit de corps," "local patriotism," "nationalism," etc. The integrative tendencies, on the other hand, are manifested in "co-operativeness," "disciplined behaviour," "loyalty," "self-effacement," "devotion to duty," "internationalism," and so on.

Note, however, that most of the terms referring to higher levels of the hierarchy are ambiguous. The loyalty of individuals towards their clan reflects their integrative tendencies; but it enables the clan as a whole to behave in an aggressive, self-assertive way. The obedience and devotion to duty of the members of the Nazi S.S. Guard kept the gas chambers going. "Patriotism" is the virtue of subordinating private interests to the higher interests of the nation; "nationalism" is a synonym for the militant expression of those higher interests. The infernal dialectic of this process is reflected throughout human history. It is not accidental; the disposition towards such disturbances is inherent in the part-whole polarisation of social hierarchies. It may be the unconscious reason why the Romans gave the god Janus such a prominent role in the Pantheon as the keeper of doorways, facing both inward and outward, and why they named the first month of the year after him.

The Integrative Powers of Life

Let us go for a moment to the organelles which operate inside the cell. The mitochondria transform food—glucose, fat, proteins—into the

The interior of the Pantheon (Giovanni Battista Piranesi, 1768). The god Janus, guardian of entrances, was sculpted into the doorways of many Roman buildings, including the magnificently domed Pantheon, which is two thousand years old.

chemical substance adrenosine-triphosphate, ATP for short, which all animal cells utilise as fuel. It is the only type of fuel used throughout the animal kingdom to provide the necessary energy for muscle cells, nerve cells and so on; and there is only this one type of organelle throughout the animal kingdom which produces it. The mitochondria have been called "the power plants of all life on earth." Moreover, each mitochodrion carries not only its set of instructions how to make ATP, but also its own hereditary blueprint, which enables it to reproduce itself independently from the reproduction of the cell as a whole.

Until a few years ago, it was thought that the only carriers of heredity were the chromosomes in the nucleus of the cell. At present we know that the mitochondria, and also some other organelles located in the cytoplasm (the fluid surrounding the nucleus) are equipped with their own genetic apparatus, which enables them to reproduce independently. In view of this, it has been suggested that these organelles may have evolved independently from each other at the dawn of life on this planet, but at a later stage had entered into a kind of symbiosis.

This plausible hypothesis sounds like another version of the watchmaker's parable; we may regard the stepwise building up of complex hierarchies out of simpler holons as a basic manifestation of the integrative tendency of living matter. It seems indeed very likely that the single cell, once considered the atom of life, originated in the coming together of molecular structures which were the primitive forerunners of the organelles, and which had come into existence independently, each endowed with a different characteristic property of life—such as self-replication, metabolism, motility. When they entered into symbiotic partnership, the emergent whole proved to be an incomparably more stable, versatile and adaptable entity than a mere summation of the parts would imply. To quote Ruth Sager:

> Life began, I would speculate, with the emergence of a stabilised tripartite system: nucleic acids for replication, a photosynthetic or chemosynthetic system for energy conversion, and protein enzymes to catalyse the two processes. Such a tripartite system could have been the ancestor of chloroplasts and mitochondria and perhaps of the cell itself. In the course of evolution, these primitive systems might have coalesced into the larger framework of the cell.

The hypothesis is in keeping with all we know about that ubiquitous manifestation of the integrative tendency: symbiosis, the varied forms of partnership between organisms. It ranges from the mutually indispensable association of algae and fungi in lichens, to the less intimate but no less vital inter-dependence of animals, plants and bacteria in ecological communities.

Where different species are involved, the partnership may take the form of "commensalism"—barnacles travelling on the sides of a whale; or of "mutualism," as between flowering plant and pollinating insects, or between ants and aphids—a kind of insect "cattle" which the ants protect and "milk" for their secretions in return. Equally varied are the forms of co-operation within the same species, from colonial animals upward. The Portuguese man-of-war is a colony of polyps, each specialised for a particular function; but to decide whether its tentacles, floats and reproductive units are individual animals or mere organs is a matter of semantics; every polyp is a holon, combining the characteristics of independent wholes and dependent parts.

The same dilemma confronts us, on a higher turn of the spiral, in the insect societies of ants, bees, termites. Social insects are physically separate entities, but none can survive if separated from its group; their existence is completely controlled by the interests of the group as a whole; all members of the group are descendents from the same pair of parents, interchangeable and indistinguishable, not only to the human eye but also probably to the insects themselves, which are supposed to recognise members of their group by their smell, but not to discriminate between individuals.

On Individuality

An individual is usually defined as an indivisible, self-contained unit, with a separate, independent existence of its own. But individuals in this absolute sense are nowhere found in Nature or society, just as we nowhere find absolute wholes. Instead of separateness and independence, there is co-operation and interdependence, running through the whole gamut, from physical symbiosis to the cohesive bonds of the swarm, hive, shoal, flock, herd, family, society.

The picture becomes even more blurred when we consider the criterion of "indivisibility." The word "individual" originally means just that; it is derived from the Latin "in-dividuus"—as atom is derived from the Greek "a-tomos." But on every level, indivisibility turns out to be a relative affair. Protozoa, sponges, hydra and flatworms can multiply by simple fission or budding: that is, by the breaking up of one individual into two or more, and so on. As Ludwig von Bertalanffy wrote: "How can we call these creatures individuals when they are in fact 'dividua,' and their multiplication arises precisely from division? Can we insist on calling a hydra or a turbelerian flatworm an individual, when these animals can be cut into as many pieces as we like, each capable of growing into a complete organism?" A flatworm, cut into six slices, will actually regenerate a complete individual from each slice within a matter of weeks. If the wheel of rebirth transforms me into a flatworm meeting a similar fate, must I then assume that my immortal soul has split into six immortal solons?

Christian theologians will find an easy way out of this dilemma by denying that animals have souls; but Hindus and Buddhists take a different view. And secular-minded philosophers, who do not talk about souls, but affirm the existence of a conscious ego, also refuse to draw a boundary line between creatures with and without consciousness. But if we assume that there exists a continuous scale of gradations, from the sentience of primitive creatures, through various degrees of consciousness, to full self-awareness, then the experimental biologist's challenge to the concept of individuality poses a genuine dilemma. The only solution seems to be to get away from the concept of the individual as a monolithic structure, and to replace it by the concept of the individual as an open hierarchy whose apex is forever receding, striving towards a state of complete integration which is never achieved.

The regeneration of a complete individual from a small fragment of a primitive animal is an impressive manifestation of the integrative powers of living matter. But there are even more striking examples. Nearly a generation ago, Wilson and Child showed that if the tissues of a living sponge—or hydra—are crushed to pulp, passed through a fine filter, and the pulp is then poured into water, the dissociated cells will soon begin to associate, to aggregate first into flat sheets, then round up into a sphere, differentiate progressively and end up as adult individuals with characteristic mouth, tentacles and so forth. More recently, Paul Weiss and his associates have demonstrated that the developing organs in animal embryos are also capable, just like sponges, of re-forming, after having been pulped.

Weiss and James cut out bits of tissue from eight- to fourteen-day-old chick embryos, minced and filtered the tissues through nylon sheets, recompacted them by centrifuging, and transplanted them to the membrane of another growing embryo. After nine days, the scrambled liver cells had started forming a liver, the kidney cells a kidney, the skin cells to form feathers. More than that: the experimenters were also able to produce normal embryonic kidneys by mincing, pooling and scrambling kidney tissues from several different embryos. The holistic properties of these tissues survived not only disintegration but also fusion.

In the light of such experimental data, the homely concept of the individual vanishes in the mist. If the crushed and re-formed sponge possesses individuality, so does the embryonic kidney. From organelles to organs, from organisms living in symbiosis to societies with more complex forms of interdependence, we nowhere find completely self-contained wholes, only holons—double faced entities which display the characteristics both of independent units and of inter-dependent parts.

The Self-Assertive Tendency

In the previous pages I have emphasized the phenomena of interdependence and partnership, the integrative potential of holons to behave as parts of a more complex whole. The other side of the story reveals, instead of cooperation, competition between the parts of the whole, reflecting the self-assertive tendency of holons on every level. Even plants, which are mostly green and not "red in tooth and claw," compete for light, water and soil. Animal species compete with each other for ecological niches, predator and prey compete for survival, and within each species there is competition for territory, food, mates and dominance.

There is also a less obvious competition between holons within the organism in times of stress, when the exposed or traumatised parts tend to assert themselves to the detriment of the whole. Under normal conditions, however, when the organism or body social is functioning steadily, the integrative and self-assertive tendencies are in a state of dynamic equilibrium—symbolised by Janus Patulcius, the Opener, with a key in his left hand, and Janus Clusius, the Closer, jealous guardian of the gate, with a staff in his right. To sum up, the living organism is not a mosaic of aggregate elementary physico-chemical processes, but a hierarchy of parts with parts, in which each holon, from the sub-cellular organelles upward, is a closely integrated structure, equipped with self-regulatory devices, and enjoys a degree of self-government. Transplant surgery and experimental embryology provide striking illustrations for the autonomy of organismic holons.

The integrative powers of life are manifested in the phenomena of symbiosis between organelles, in the varied forms of partnership within the same species or between different species; in the phenomena of regeneration, in lower species, of complete individuals from their fragments; in the re-formation of scrambled embryonic organs, etc. The self-assertive tendency is equally ubiquitous in the competitive struggle for life.

Why Hierarchy Theory?

Science is only just beginning to rid itself of the mechanistic preconceptions of the nineteenth century—the world as a billiard table of colliding atoms—and to realize that hierarchic organization is a fundamental principle of living nature; that it is "the essential and distinguishing characteristic of life" (Howard Pattee); and that it is "a real phenomenon, presented to us by the biological object, and not the fiction of a speculative mind" (Paul Weiss). It is at the same time a conceptual tool which on some occasions acts as an Open Sesame.

The universal applicability of the hierarchic model may arouse the suspicion that it is logically empty. I hope to show that this is not the case, and that the search for the fundamental properties, or laws, which all these varied hierarchies have in common amounts to more than a play on superficial analogies—or to riding a hobby horse. It should rather be called an exercise in General Systems Theory—that relatively recent interdisciplinary school, founded by von Bertalanffy, whose purpose is to construct theoretical models and discover general principles which are universally applicable to biological, social and symbolic systems of any kind. As early as 1936, Joseph Needham wrote: "The hierarchy of relations, from the molecular structure of carbon compounds to the equilibrium of species and ecological wholes, will perhaps be the leading idea of the future." Unfortunately, the term "hierarchy" itself is rather unattractive and often provokes a strong emotional resistance. It is loaded with military and ecclesiastic associations, or evokes the "pecking hierarchy" of the barnyard, and thus conveys the impression of a rigid, authoritarian structure, whereas in the present theory a hierarchy consists of autonomous, self-governing holons endowed with varying degrees of flexibility and freedom. Encouraged by the friendly reception of the holon, I shall occasionally use the terms "holarchic" and "holarchy," but without undue insistence.

Every holon is possessed of two opposite tendencies or potentials: an integrative tendency to function as part of the larger whole, and a self-assertive tendency to preserve its individual autonomy. The postulate of a universal self-assertive tendency needs no apology; it has an immediate appeal to common sense, and has many forerunners—such as the "instinct for self-preservation," "survival of the fittest," and so forth. But to postulate as its counterpart an equally universal integrative tendency, and the dynamic interplay between the two as the key to a general systems theory, smacks of old-fashioned vitalism and runs counter to the Zeitgeist, epitomized in books like Jacques Monod's *Chance and Necessity* or B. F. Skinner's *Beyond Freedom and Dignity*.

It may therefore be appropriate to wind up this chapter with a few quotations from a recent book by an eminent clinician, Dr. Lewis Thomas (President of the Sloan-Kettering Cancer Centre), who can hardly be accused of an unscientific attitude. The passage starts with a fascinating description of the parasite *Myxotricha paradoxa*, a single-celled creature which inhabits the digestive tract of Australian termites.

> At first glance, he appears to be an ordinary, motile protozoan, remarkable chiefly for the speed and directness with which he swims from place to place, engulfing fragments of wood finely chewed by his termite host. In the termite ecosystem, an arrangement of Byzantine complexity, he stands at the epicenter. Without him, the wood, however finely chewed, would never get digested; he supplies the enzymes that break down cellulose to edible carbohydrate, leaving only the nondegradable lignin, which the termite then excretes in geometrically tidy pellets and uses as building blocks for the erection of arches and vaults in the termite nest. Without him there would be no termites, no farms of the fungi that are cultivated by termites and will grow nowhere else.

But this tiny creature inside the termite's digestive tract turns out to consist of whole populations of even tinier creatures living in symbiosis with each other, yet retaining the autonomous individuality. Thus "the flagellae that beat in synchrony to propel myxotricha with such directness turn out, on closer scrutiny with the electron microscope, not to be flagellae at all. They are outsiders, in to help with the business: fully formed perfect spirochetes that have attached themselves at regularly spaced intervals all over the surface of the protozoan." Thomas then

In the late seventeenth century, Marcello Malpighi conducted a detailed study of plant anatomy, which yielded this exquisite drawing of the root of a pea plant. Malpighi wrongly surmised that the nodules were pathological. Two centuries later the chemist Marcellin Bethelot showed that these nodules were really symbiotic structures for capturing a vital nutrient, nitrogen, from the atmosphere.

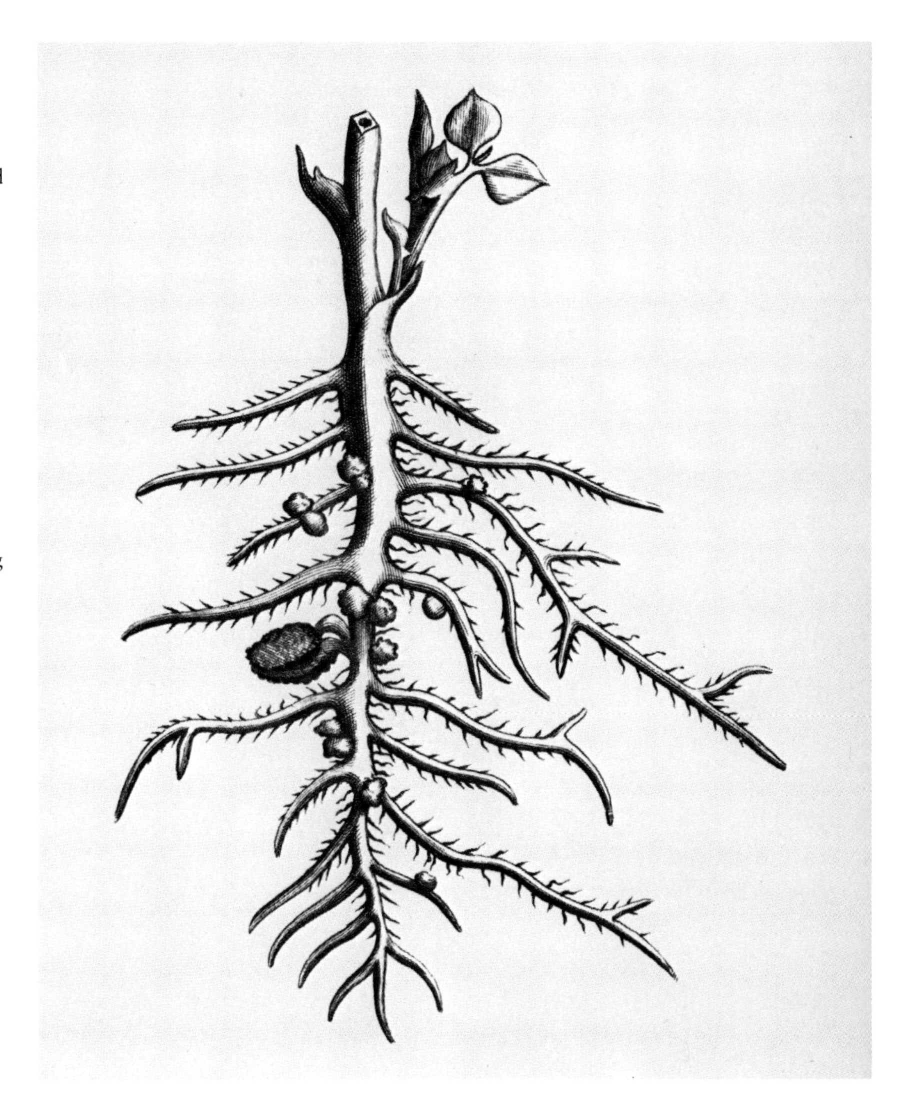

enumerates the various types of other organelles and bacteria which form a kind of cooperative zoo inside *Myxotricha*, and cites evidence that the cells which constitute the human body evolved by a similar process "of being made up, part by part, by the coming together of just such prokaryotic animals." Thus the lowly *Myxotricha* becomes a paradigm for our integrative tendency.

> The whole animal, or ecosystem, stuck for the time being halfway along in evolution, appears to be a model for the development of cells like our own. . . . There is an underlying force that drives together the several creatures comprising myxotricha, and then drives the assemblage into union with the termite. If we could understand this tendency, we would catch a glimpse of the process that brought single separate cells together for the construction of metazoans, culminating in the invention of roses, dolphins, and, of course, ourselves. It might turn out that the same tendency underlies the joining of organisms into communities, communities into ecosystems, and ecosystems into the biosphere. If this is, in fact, the drift of things, the way of the world, we may come to view immune reactions, genes for the chemical marking of self, and perhaps all reflexive responses of aggression and defense as secondary developments in evolution, necessary for the regulation and modulation of symbiosis, not designed to break into the process, only to keep it from getting out of hand.
>
> If it is in the nature of living things to pool resources, to fuse when possible, we would have a new way of accounting for the progressive enrichment and complexity of form in living things.

The Evolutionary Journey

Evolution has been compared to a journey from an unknown origin towards an unknown destination, a sailing along a vast ocean; but we can at least chart the route which carried us from the sea-cucumber stage to the conquest of the moon; and there is no denying that there is a wind which makes the sails move. But whether we say that the wind, coming from the distant past, pushes the boat along, or whether we say that it drags us along into the future, is a matter of choice. The purposiveness of all vital processes, the strategy of the genes and the power of the exploratory drive in animal and man, all seem to indicate that the pull of the future is as real as the pressure of the past. Causality and finality are complementary principles in the sciences of life; if you take out finality and purpose you have taken the life out of biology as well as psychology. If this be called vitalism, I have no objection, and shall quote in reply a profound remark by that arch-vitalist, Henri Bergson: "The vitalist principle may indeed not explain much, but it is at least a sort of label affixed to our ignorance, so as to remind us of this occasionally, while mechanism invites us to ignore that ignorance."

Arthur Koestler's inclusion of passages by Henri Bergson, Lewis Thomas, and others transforms this anthology into a holarchy itself. I suspect that if Bios the watchmaker had a mind to communicate humanity's great ideas about his namesake, he would choose the same technique as have I.

In the previous essay Arthur Koestler referred to Ludwig von Bertalanffy as the founder of general systems theory, "whose purpose is to construct theoretical models and discover general principles which are universally applicable to biological, social, and symbolic systems of any kind." Traditionally, scientists have searched for principles of physics that can explain chemistry, chemical principles that can explain biology, and biological principles that can explain psychology. The goal is to reduce chemistry to physics, biology to chemistry, and psychology and even sociology to biology. Systems theorists, on the other hand, look for general principles that apply across the natural and human sciences.

The principles themselves, like Koestler's concept of holons, stand outside any single field. Holons are not, for example, chemical entities to which cats and caterpillars can be reduced. Rather, holons are an abstract idea that can provide insight into the workings of many complex systems. Systems theorists, moreover, believe that reductionist methods are inadequate to the task of understanding organized complex systems.

The Realm of Organized Complexity

What is an organized complex system? A living organism is the best example. Indeed, Ludwig von Bertalanffy got the notion that reductionism is not the path to all answers while he was working in the field of biology. But what exactly is an organized complex system? The figure below provides a visual answer. It is based on a diagram that appeared in An Introduction to General Systems Thinking *by Gerald Weinberg, copyright 1975 by John Wiley & Sons, Inc., reprinted here by permission of John Wiley & Sons. Weinberg explains that diagram thus:*

Gerald Weinberg

Region 1 is the region that might be called "organized simplicity"—the region of machines or mechanisms. Region 2 is the region of "unorganized complexity"—the region of populations, or aggregates, as we shall call them. Region 3, the yawning gap in the middle, is the region of "organized complexity"—the region too complex for analysis and too organized for statistics. This is the region of systems.

Region 2
Unorganized complexity
(aggregates)

Region 3
Organized complexity
(systems, life)

Region 1
Organized simplicity
(crystals, machines)

Weinberg's book is punctuated with striking aphorisms about the systems way of thinking. These statements sound remarkably similar to some of the ideas expressed by James Lovelock, who envisioned the grandest of all complex systems, Gaia. For example, Weinberg declares,

- To be a successful generalist, one must study the art of ignoring data and of seeing only the mere outline of things.

- To be a successful generalist, we must approach complex systems with a certain naive simplicity.

- To be a good generalist, one should not have faith in anything. Faith, as Bertrand Russell once pointed out, is the belief in something for which there is no evidence. Every article of faith is a restriction on the free movement of thought.

- The willingness to make a fool of oneself should almost be a requirement for admission to the Society for General Systems Research.

Weinberg expresses all these ideas in an anecdote: Okubo Shibutsu, famous for painting bamboo, was requested to execute a kakemono representing a bamboo forest. Consenting, he painted with all his known skill a picture in which the entire bamboo grove was in red. The patron upon its receipt marveled at the extraordinary skill with which the painting had been executed, and repairing to the artist's residence, he said: "Master, I have come to thank you for the picture; but excuse me, you have painted the bamboo in red." "Well," cried the master, "in what color would you desire it?" "In black, of course," replied the patron. "And who," answered the artist, "ever saw a black-leaved bamboo?"

The systems approach in biology is well stated by Stephen Jay Gould in an essay that appeared in the January 1984 issue of Natural History. *While chronicalling the scientific career of Ernest Everett Just, an early practitioner of holism in biological research whose career at the turn of the century was held back by racial prejudice (Just was black), Gould identifies three main elements of a nonreductionist approach to biology.*

Stephen Jay Gould

First, nothing in biology contradicts the laws of physics and chemistry; any adequate biology must be consonant with the "basic" sciences. Second, the principles of physics and chemistry are not sufficient to explain complex biological objects because new properties emerge as a result of organization and interaction. These properties can only be understood by the direct study of whole, living systems in their normal state. Third, the insufficiency of physics and chemistry to encompass life records no mystical addition, no contradiction to the basic sciences, but only reflects the hierarchy of natural objects and the principle of emergent properties at higher levels of organization.

Bamboo, section of a handscroll (Hsü Wei, sixteenth century)

Although Just and other research biologists were practicing a non-reductionist science before Ludwig von Bertalanffy entered the field, von Bertalanffy was the first in the West to fully articulate this approach. His Problems of Life *(copyright 1952 by C. A. Watts & Co., Ltd.) is still perhaps the most engaging and profound presentation of a method and view of biology that transcends the mechanist-vitalist dispute. (Excerpts are reprinted here by permission of A. Francke Co., Bern, Switzerland.)*

Biology as the Central Science

Ludwig von Bertalanffy

The right to speak of a biological worldview follows from the central position occupied by biology in the hierarchy of the sciences. Biology is based on physics and chemistry, the laws of which are an indispensable groundwork for the investigation and explanation of the phenomena of life. It embraces an abundance of particular problems—such, for example, as those of organic form, purposiveness, and phylogenetic evolution—that are alien to physics and make the biologist's research and concepts different from those of the physicist. Finally, biology provides the basis of psychology and sociology; for the investigation of mental activity is based upon its physiological foundations, and similarly the theory of human relations cannot neglect their biological bases and laws. Because of this central position among the sciences, biology has possibly the greatest multiplicity of problems: The phenomenon of "life" is a meeting-place for those conceptions which, according to the usual distinction, originate in the exact sciences on the one hand and in the social sciences on the other.

But the purport of biology for modern intellectual life is even more deeply rooted. The world-concept of the nineteenth century was a physical one. Physical theory, as it was then understood—a play of atoms controlled by the laws of mechanics—seemed to indicate the ultimate reality underlying the worlds of matter, life, and mind, and it provided the ideational model also for the non-physical realms, the living organism, mind, and human society. Today, however, all sciences are beset by problems which are indicated by notions such as "wholeness," "organization," or "gestalt"—concepts that have their root in the biological field.

In this sense biology has an essential contribution to make to the modern world-concept. True, it took things rather easily in the past, when it adopted its basic conceptions from other sciences. It borrowed the mechanistic view from physics, vitalism from psychology and selection from sociology. But its mission, both as a science conceiving and mastering the phenomena peculiar to its own field and as contributing to our basic conception of the world, will be accomplished only by its autonomous development. This is the significance of the striving after new conceptions that has taken place in biology in the past decades.

For more than twenty years the author has been advocating a biological standpoint known as the "organismic conception." We shall see that biology is an autonomous science in the sense that its problems require the development of specific conceptions and laws; further, that biological knowledge and conceptions are active in different fields. In the present volume we give a survey of basic biological problems and laws within the framework of the organismic conception. From there we proceed to questions of biological knowledge, and eventually arrive at the general principles of the modern conception of the world and the claim for a "General System Theory."

What Is Life?

At a time of tremendous upheaval comparable with what we have today, science was presented with an idea that was to influence profoundly man's conception of the world. The time was the Thirty Years' War, and the man who expressed the idea was the French philosopher René Descartes. Impressed by the successes achieved in the young science of physics, then in the throes of its first progress and foreshadowing the possibilities since realized in modern technology, Descartes formulated his theory of the "bête machine." Not only did the inanimate world obey the laws of physics—this was how Descartes's thoughts went—but so also did all living organisms. Descartes therefore interpreted animals as machines, of a very complicated kind, to be sure, but comparable in principle nevertheless with man-made machines, the actions of which are governed by the laws of physics.

True, Descartes was not wholly consistent. A faithful son of the Church, he set a limit to the knowledge of physics: Man was not to be regarded as a mere machine, but as endowed with free will not submitted to the law of nature. *(Hence the Cartesian dualism of mind and matter.)* Even this limitation was to be overcome by French Enlightenment. In 1748 the Chevalier Julien de la Mettrie set up the "homme machine" against the "bête machine" of Descartes.

These thinkers sought the answer to one of the age-old problems of philosophy. A living organism, plant or animal, is apparently very different from non-living things, such as crystals, molecules, or planetary systems. Life is expressed in an endless variety of plant and animal forms. These exhibit a unique organization proceeding from the single cell to tissues, organs, and multicellular organisms composed of myriads of cells. The life processes are equally unique. Every living thing maintains itself in a continuous exchange of composing materials and energies. It can respond to external influences, the so-called stimuli, with activities, and especially with movements. Indeed, it frequently shows movements and other activities without any stimulus from outside, and in this we have an obvious, though by no means decisive, contrast between non-living and living things, in that the former are set in motion only by external forces, while the latter can show "spontaneous" movements. Organisms go through progressive transformations, which we call growth, development, senescence, and death. They are produced only by their kind by the process known as reproduction. In general, the offspring resemble the parents, a phenomenon we call heredity.

A survey of the organic world shows, however, that it represents a stream of forms surging up through geological time. These forms appear to be related by reproduction and evolution, changes having occurred in the course of ages, leading to the efflorescence of higher forms from the lower forms. Organic structures and functions are admirably fitted for the "purposes" they serve. An astounding multiplicity of processes goes on even in the simplest cell, so arranged that its identity is maintained in this ceaseless and tremendously complicated play. Equally, every living being displays in its organs and functions a purposeful construction, adapted to the environment in which it normally exists.

If the peculiar nature of living organisms is thus evident—and we hardly ever find ourselves in doubt whether we have a living thing or an inanimate one in front of us—then the question must arise whether or not an intrinsic distinction really exists between the realm of the living and that of the non-living. We ourselves are living beings, so the answer to this question must determine, in large measure, the place we assign to man.

The application of the laws and methods of the physical sciences to the phenomena of life has led to an uninterrupted series of triumphs, both in theoretical knowledge and in the practical control of nature. Descartes initiated the school of physicians and physiologists known to the history of science as the iatromechanics, who tried to explain the function of muscles and bones, the movements of blood, and similar phenomena on the basis of mechanical principles. Harvey's discovery of the circulation of the blood (1628) marked the beginning of modern physiology. Later on the application of acoustics and optics, of the theory of electricity, heat theory, energetics, and other physical fields provided an inexhaustible source of knowledge and helped to explain an ever-increasing number of biological phenomena.

Biophysics was augmented by biochemistry. Once it was believed that organic compounds, which are characteristic of living beings, and are, in nature, found exclusively in them, could be produced only in the life-processes. In the year 1828, however, Wöhler produced urea in the laboratory, the first organic compound to be synthesized. Since then, organic chemistry and biochemistry have become most important fields in modern science. They also form the basis of the chemical industry, from the chemistry of dyes to the hydrogenation of coal, the manufacture of artificial rubber *(and now plastics)*, and the therapeutic armamentarium of modern medicine, including vitamins, hormones, and the chemotherapy of today.

With the publication of Darwin's *Origin of Species* in 1859, the theory of evolution triumphed. Whereas the great systematist Linnaeus had considered animals and plant species as the work of individual acts of the Creator, now an enormous array of facts in all biological fields has been collected to demonstrate that the organic world has climbed, through long generations and geological times, from lower and simpler forms to higher and more complicated ones. At the same time, with his theory of natural selection, Darwin propounded an explanation for this evolution. Every now and then small accidental variations appear in species. They may be disadvantageous, indifferent, or useful. If disadvantageous, they are soon eliminated by natural selection in the struggle for existence; if, however, they happen to be useful, they give their owners an advantage in the competition of life so that they are more likely to survive and reproduce their kind. Thus, in the course of generations, useful variations are preserved and enhanced. Repeated through long ages, this process has led to the evolution of the different forms of living organisms and their progressive adaptations to their environments. Whereas Descartes had pointed to a divine Creator as the engineer of living machines, now the origin of purposiveness in the living world seemed explained on the basis of chance variations and selections, eliminating all purposive agents.

Hans Driesch and His Sea Urchins

Thus, the programme put forward by Descartes was the starting-point for developments that not only form the basis of biological science, but also exert a profound influence on human life. Yet in spite of these triumphs, the suspicion has never quite died down that perhaps the very essentials of life have remained untouched and unexplained. Just one year after La Mettrie's *Homme Machine,* a polemic pamphlet with the title "Man Not a Machine" was printed in London. The story goes that the author was none other than La Mettrie himself. Should this be true, the chevalier has given evidence of a sovereign irony and freedom of mind that is almost unique in the history of science.

In the course of time, this antagonistic viewpoint has been expressed in many different ways. The form most important even now, because it is logically the most consistent, is that given by Hans Driesch (since 1893). Driesch was one of the founders of developmental mechanics, that branch of biology which has as its subject the experimental investigation of embryonic development, and a classic experiment led him to reject the physico-chemical theory of life:

In the greenish depths of the sea, the sea-urchins lead a contemplative life, aloof from the problems of world and science. Yet these same peaceful creatures became the cause of a long-drawn and violent controversy about the essence of life. When a sea-urchin egg begins to develop, it divides first into two, then into four, eight, sixteen, and finally many cells, and in a series of characteristic stages it eventually forms a larva looking somewhat like a spiked helmet and known to science as a pluteus; from this the sea-urchin finally develops by way of a complicated metamorphosis. Driesch divided a sea-urchin germ, just at the beginning of its development, into two halves. One would expect that from such a half-germ only half an animal would develop. In fact, however, the experimenter watches a ghostly performance like that in Goethe's *Sorcerer's Apprentice*: "Wehe, wehe, beide Teile stehn in Eile schon als Knechte völlig fertig in die Höhe"—out of each half comes not a half but a whole sea-urchin larva, a bit smaller, it is true, but normal and complete.

The production of whole organisms from divided germs is possible with many other animals. Even identical twins, which occasionally appear in man, are produced in a similar way; they are, so to speak, a Driesch experiment performed by nature herself. The reverse experiment and other arrangements are also possible. Under certain conditions two united germs produce a unitary giant larva; by pressing an embryo between glass plates the arrangement of cells can be severely altered, and still normal larvae are produced.

Like the Sorcerer's Apprentice, Driesch found something uncanny in his experiment, and he came to the conclusion that physical laws of nature are transgressed here. Supposing that only physical and chemical forces operate in the germ, the arrangement of processes which eventually leads to the formation of an organism can be explained, according to Driesch, only by assuming that the processes are directed in the right way by means of a fixed structure, a "machine" in the widest sense of the word. But there cannot be a machine in the germ; for a machine cannot achieve the same performance, in this case the production of a normal organism, when it is divided, when its parts are dislocated, or when two complete machines are fused.

Thus, Driesch stated here the physico-chemical explanation of life reaches its limit, and only one interpretation is possible. In the embryo, and similarly in other vital phenomena, a factor is active which is fundamentally different from all physico-chemical forces, and which directs events in anticipation of the goal. This factor, which "carries the goal within itself," namely, the production of a typical organism in normal as well as experimentally disturbed development, was called entelechy by Driesch, using an Aristotelian notion. Looking around for purposefully acting factors, we find their like in our own intentional actions. It is factors that are ultimately comparable with the mental factors in our purposive actions which make the crucial difference between the living and the non-living and which cause the more than mechanical and physical properties of life. In this way we find two fundamental and antithetical biological conceptions, which in their beginnings go back to the dawn of Greek philosophy. They are customarily termed "mechanism" and "vitalism."

Mechanism versus Vitalism

The expression "mechanistic theory" has been used in widely different senses, a fact that has much encumbered and confused the issue. We have already mentioned the two most important meanings of this term. First, the mechanistic conception sees in living things only a complicated play of those forces and laws which are also present in inanimate nature. A second meaning is seen in the machine-theory of life; the arrangement of events characteristic of all processes in the cell and the organism is interpreted in terms of structural conditions.

In contrast, vitalism denies the possibilities of a complete physico-chemical explanation of life and maintains an intrinsic difference between the living and non-living. It starts, as we have seen in Driesch's doctrine, from the phenomena of regulation, ie., restitution after disturbances, which seems inexplicable on the basis of a machine. Other vitalists arrive at their conception by carrying through consistently the machine theory of life.

Every machine implies an engineer to design and build it. In this sense, Descartes drew a logical conclusion when he inferred a divine spirit as creator of living machines. Darwin's theory put chance in the place of the creative spirit. Modern biology has made it highly probable that this explanation holds good at least for the origin of varieties and species, and perhaps also for some of the higher systematic units. It is, however, much harder to decide whether it is also sufficient to explain the origin of the great plans of organization and the origin of that interaction of innumerable physiological processes necessary to the functioning of every organism. Locomotives and watches do not usually arise in nature by a play of chance forces—do then infinitely more complicated organic "machines"? Thus, the arrangement of the superlatively numerous physico-chemical processes, by means of which the organism is maintained and restored even after serious disturbances, and, further, the origin of the complicated "machine" of the organism cannot be explained, according to vitalistic doctrine, save by the action of specific vital factors, whether we call them Entelechy, Unconscious, or World Soul, which interfere, purposely and directively, with physico-chemical events.

At once, however, we see that vitalism must be rejected as far as scientific theory is concerned. According to it, structure and function in the organism are governed, as it were, by a host of goblins, who invent and design the organism, control its processes, and patch the machine up after injury. This gives us no deeper insight; but we merely shift what at present seems inexplicable to a yet more mysterious principle and assemble it into a question mark that is inaccessible to research.

Vitalism says nothing else than that the essential problems of life lie outside the sphere of natural science. If that were so, then scientific research would become pointless; for even with the most complicated experiments and apparatus, it can lead to no other explanation than the anthropomorphism of primitive mankind, who see an elfin intelligence and will similar to their own in the apparent directiveness and purposiveness in living nature. Whether we consider the behaviour of an animal, or the multiplicity of physical and chemical processes in a cell, or the development of organic structures and functions, we always get the same answer—it is just a soul-like something standing behind them and directing them.

The history of biology is the refutation of vitalism, for it shows that always it was just those phenomena which appeared inexplicable at the time that seemed the domain of vitalistic factors. Thus the production of organic compounds was considered a vitalistic phenomenon up to the time of Wöhler; so was the fermentative activity of the cell even for Pasteur; and until Buchner carried through fermentation with yeast extracts at the end of the nineteenth century, so was the phenomenon of organic regulation in Driesch's doctrine. But progress in research has brought an ever-increasing number of phenomena, previously regarded as vitalistic, into the realm of scientific explanation and law.

The contest between the mechanistic and vitalistic conceptions is like a game of chess played over nearly two thousand years. It is essentially the same arguments that always come back, though in manifold disguises, modifications, and forms. In the last resort, they are an expression of two opposing tendencies in the human mind. On the one hand, there is the tendency to subordinate life to scientific explanation and law; on the other hand, there is the experience of our own mind, taken as a standard for living nature, and inserted in the supposed or actual gaps in our scientific knowledge.

Introducing a Third View

In our time, a fundamental change of scientific conceptions has occurred. The revolutions in modern physics are widely known. They have led, in the relativity and quantum theories, to a radical reform and expansion of physical doctrine, outranking the progress made in centuries of the past. Less obvious, but perhaps not less significant in their consequences, are the changes that have led both to a new attitude towards the basic problems of living nature and to new questions and solutions.

We might take as an established fact of the modern development in biology that it does not consent completely to either of the classical views, but transcends both in a new and third one. This attitude has been called the "organismic conception" by the author, who has worked it out for more than twenty years. Similar conceptions have been found necessary, and have been evolved, in the most diverse fields of biology, as well as in the neighbouring sciences of medicine, psychology, sociology, etc. If we retain the term "organismic conception" we shall consider it merely as a convenient denomination for an attitude which has already become very general and largely anonymous. This seems to be justified, in so far as the author was probably the first to develop the new standpoint in a scientifically and logically consistent form.

It appeared to be the goal of biological research to resolve the complex entities and processes that confront us in living nature into elementary units—to analyse them—in order to explain them by means of the juxtaposition or summation of these elementary units and processes. Procedure in classical physics supplied the pattern. Thus chemistry resolves material bodies into elementary components—molecules and atoms; physics considers a storm that tears down a tree as the sum of movements of air particles, the heat of a body as the sum of the energy of motion of molecules, and so on. A corresponding procedure was applied in all biological fields, as some examples will easily show.

Biochemistry investigates the individual chemical constituents of living bodies and the chemical processes going on within them. In this way it specifies the chemical compounds found in the cell and the organism as well as their reactions. The classical "cell theory" considered cells as the elementary units of life, comparable to atoms as the elementary units of chemical compounds. So a multicellular organism appeared morphologically as an aggregate of such building units. Genetics considered the organism as an aggregate of characters going back to a corresponding aggregate of genes in the germ cells, transmitted and acting independently of each other. Accordingly, the theory of natural selection resolved living beings into a complex of characters, some useful, others disadvantageous, which characters, or rather their corresponding genes, are transmitted independently, thus through natural selection affording the opportunity for the elimination of unfavourable characters, while allowing favourable ones to survive and accumulate.

The same principle could be shown to operate in every field of biology, and in medicine, psychology, and sociology as well. The examples given will suffice, however, to show that the principle of analysis and summation has been directive of all fields. Analysis of the individual parts and processes in living things is necessary, and is the prerequisite for all deeper understanding. Taken alone, however, analysis is not sufficient.

The phenomena of life—metabolism, irritability, reproduction, development, and so on—are found exclusively in natural bodies which are circumscribed in space and time, and show a more or less complicated structure; bodies that we call organisms. Every organism represents a "system," by which term we mean a complex of elements in mutual interaction. From this obvious statement the limitations of the analytical and summative conceptions must follow. First, it is impossible to resolve the phenomena of life completely into elementary units; for each individual part and each individual event depends not only on conditions within itself, but also to a greater or lesser extent on the conditions within the whole, or within superordinate units of which it is a part. Hence the behavior of an isolated part is, in general, different from its behaviour within the context of the whole.

Secondly, the actual whole shows properties that are absent from its isolated parts. The problem of life is that of organization. As long as we single out individual phenomena we do not discover any fundamental difference between the living and the non-living. Certainly organic molecules are more complicated than inorganic ones; but they are not distinguishable from dead compounds by fundamental differences. Even complicated processes considered a long time as being specifically vital, like those of cell respiration and fermentation, morphogenesis, nerve action, and so on, have been explained to a large extent physicochemically, and many of them can even be imitated in inanimate models.

A fundamentally new problem is presented, however, in the singular and specific arrangement of parts and processes that we meet with in living systems. Even a knowledge of all the chemical compounds that build a cell would not explain the phenomena of life. Already the simplest cell is a superlatively complex organization, the laws of which are at present only dimly seen. A "living substance" has been spoken of.

This concept is due to a fundamental fallacy. There is no "living substance" in the sense that lead, water, or cellulose are substances, where any arbitrarily taken part shows the same properties as the rest. Rather is life bound to individualized and organized systems, the destruction of which puts an end to it.

Similar considerations apply to the processes of life. So long as we consider the individual chemical reactions that take place in a living organism we are unable to indicate any basic difference between them and those that go on in inanimate things or in a decaying corpse. But a fundamental contrast is found when we consider, not single processes, but their totality within an organism or a partial system of it, such as a cell or an organ. Then we find that all parts and processes are so ordered that they guarantee the maintenance, construction, restitution, and reproduction of organic systems. This order basically distinguishes events in a living organism from reactions taking place in non-living systems or in a corpse.

The task of biology, therefore, is to establish the laws governing order and organization within the living. Moreover, these laws are to be investigated at all levels of biological organization—at the physico-chemical level, at the level of the cell and of the multicellular organization, and finally at the level of communities consisting of many individual organisms. *(And, as Jim Lovelock would urge, at the global level of the biosphere.)* In this way the autonomy of life, denied in the mechanistic conception, and remaining a metaphysical question mark in vitalism, appears, in the organismic conception, as a problem accessible to science and, in fact, already under investigation.

Systems Theory as a Research Tool

The term "wholeness" has been much misused in past years. Within the organismic conception it means neither a mysterious entity nor a refuge for our ignorance, but a fact that can and must be dealt with by scientific methods. The organismic conception is not a compromise, a muddling through or mid-course between the mechanistic and vitalistic views. Organization and wholeness considered as principles of order, immanent to organic systems, and accessible to scientific investigation, involve a basically new attitude.

What occurred to the organismic conception was, however, what usually happens to new ideas: first it was attacked and refused, then declared to be old and self-evident. In fact, once it is realized, this conception merely draws the consequences from the obvious statement that organisms are organized. To achieve this unbiased approach it was necessary, however, and in many fields is still necessary, even today, to combat deeply rooted habits of thought.

The future historian of our times will note as a remarkable phenomenon that, since the time of the First World War, similar conceptions about nature, mind, life, and society arose independently, not only in different sciences but also in different countries. Everywhere we find the same leading motifs: the concepts of organization showing new characteristics and laws at each level, those of the dynamic nature of, and the antitheses within reality. *At this point von Bertalanffy launches into a review of the thinkers, ancient and modern, who espoused similar views. Among them are Heraclitus, Cardinal Nicholas of Cusa, Leibniz, Goethe, Alfred North Whitehead, Hegel, Karl Marx, Frederick Engels, and the twentieth century biologist J. B. S. Haldane.*

The fact that from absolutely different and even diametrically opposed starting-points, from the most varied fields of scientific research, from idealistic and materialistic philosophies, in different countries and social environments, essentially similar conceptions have evolved, shows their intrinsic necessity. It can mean nothing else than that these common general concepts are essentially true and unavoidable. A stupendous perspective emerges, a vista towards a hitherto unsuspected unity of the conception of the world. Similar general principles have evolved everywhere, whether we are dealing with inanimate things, organisms, mental or social processes. What is the origin of these correspondences?

We answer this question by the claim for a new realm of science, which we call General System Theory. It is a logico-mathematical field, the subject matter of which is the formulation and derivation of those principles which hold for systems in general. A "system" can be defined as a complex of elements standing in interaction. There are general principles holding for systems, irrespective of the nature of the component elements and of the relations or forces between them.

The significance of a general system theory lies in various directions. For example, we may distinguish various levels in the description of phenomena. The first is represented by mere analogies, that is, superficial similarities. A second level is given in logical homologies; here phenomena differ in the causal factors involved, but are governed by structurally identical laws. Analogies are scientifically worthless. Homologies, however, frequently provide very useful models and are widely used in this way in physics.

General system theory can serve therefore as a tool to distinguish analogies from homologies, to lead to legitimate conceptual models and transfer of laws from one realm to another and to prevent deceptive and inadmissible analogies. In sciences that are not within the framework of physico-chemical laws, such as demography and sociology, as well as wide fields in biology, nevertheless, exact laws can be stated if suitable model-conceptions are chosen. Logical homologies result from general system characters, and this is the reason why structurally similar principles appear in different fields, and so give rise to a parallel evolution in different sciences.

The position of general system theory is similar to that of the theory of probability, which is purely formal in itself, but can be applied to very different fields, such as theory of heat, biology, practical statistics, and so on. General system theory may be considered as a step towards that "Mathesis universalis" which Leibniz dreamed of—a comprehensive semantic system including the various sciences. Perhaps it can be said that in the modern dynamic conception a theory of systems may play a role similar to that of Aristotelian logic in antiquity. For the latter, classification was the basic attitude. In modern science, dynamic interaction is the basic problem in all fields, and its general principles will have to be formulated in general system theory.

"Life is not light, but the refracted colour"

There is still a final question that we have to answer. We have pinned down biology, in the organismic conception, at the level of pure science. We claimed that the phenomena of life are accessible to exact laws, though we may still be far from having reached this goal. We emphasized that intervention of vitalistic agents must be rejected. The question then arises: Does this mean a bleak materialism, a soulless and godless nature?

A glance at the most exact sciences answers this question. In a sweeping synthesis, physics has come to a worldview that comprehends reality from the incomprehensibly small units in the field of quanta up to the incomprehensibly large systems of galaxies. This control of nature, conceptual in physical theory and practical in technology, rests on the fact that phenomena are caught in a cobweb of logico-mathematical relations—which we call laws of nature. It is the triumph of modern physics that this fabric of laws of nature has achieved a universality and objectiveness never before reached. The trivial fact that the technological control of nature has been possible by means of those laws shows that they correspond in a wide degree to reality.

However, a certain resignation goes hand in hand with these achievements. In contrast to the self-assertion of former times, physics has realized that its task is the description of phenomena within a system of formal relations. It no longer expects to grasp the core of reality. Whereas earlier physics had thought that it found the ultimate essence in tiny hard bodies, the statements of modern physics are different. Matter is resolved into some oscillatory process—but oscillation means only a periodic change of some magnitude, the ultimate nature of which remains undecided.

The physicist does not answer the question what an electron really "is." His most penetrating insight only states the laws that are characteristic of the entity called an electron. Likewise, no answer may be expected from the biologist to the question of what life may be in its intimate essence. Even with advancing knowledge, he too will only be better able to state what laws characterize, and apply to, the phenomenon facing us as the living organism.

Factors inaccessible to objective investigation must not intrude into the laws which can be stated for the observable. On an essentially different level lies a metaphysics trying to gain an intuitive knowledge of reality. We are not only scientific intellects, we are also human beings. To express in momentous symbols the core of reality, that is what myth, poetry, and philosophy are trying.

If, however, we aspire to grasp living nature in a brief sentence, it seems that this is given in a favourite expression of Goethe's. "Dauer im Wechsel" (duration in change) it is called in a profound poem. And the river that seemed the simile of life to Heraclitus, ever changing in its waves and yet persisting in its flow, also gives final knowledge to Goethe-Faust. Incapable of looking at the sun of reality, he and the scientific mind rest content with a great metaphor, holding, however, inexhaustible powers of life and thought:

Behind me, therefore, let the sun be glowing!
The cataract, between the crags deep-riven,
I thus behold with rapture ever-glowing.
Yet how superb, across the tumult braided,
The painted rainbow's changing life is bending.
Consider, and 'tis easy understanding,
Life is not light, but the refracted colour.

In the Mountains
(Albert Bierstadt, 1867)

IV *Game Theory and the Evolution of Cooperation*

Mutual aid is as much a law of animal life as mutual struggle.
—Peter Kropotkin

The key to doing well lies not in overcoming others, but in eliciting their cooperation.
—Robert Axelrod

9 The Prisoner's Dilemma

"I never expected to find wisdom or hope for the future of our species in a computer game, but here it is, in Axelrod's book." Thus quipped Lewis Thomas in a review of Robert Axelrod's 1984 book, The Evolution of Cooperation. *Axelrod is a political scientist at the University of Michigan, but in this essay you will see how his work in game theory has led to stunning new insights about the roots of behavior in the biological as well as the social realm.*

This work recently brought Axelrod a MacArthur fellowship. The MacArthur Foundation has a most unusual program for providing financial assistance to exceptionally creative individuals in all walks of life. Like the blessings of a fairy godmother, a MacArthur fellowship is as unexpected as it is prestigious. There are no application forms.

Axelrod earned a bachelor's degree in mathematics at the University of Chicago. So how did he end up in political science? "I was looking around for some science to apply it to. Political science seemed the most interesting." Similarly, Axelrod later took the general insights derived from his work in game theory and probed their application in biology. He did so by collaborating with one of the most respected scientists in biology, William D. Hamilton of Oxford University in England.

Game theory was developed and first applied by economists and military strategists, but Axelrod, Hamilton, John Maynard Smith, and others have shown that it can be a powerful tool for understanding the evolution of organic behavior ranging from interspecies symbiosis to territoriality in birds and even to the transformation of benign, resident bacteria into pathogens if a human host becomes seriously ill. Opportunities for new insights in game theory, as well as systems theory, have been enhanced by the wide availability of high-speed computers. In both of these realms, one may seek a solution to a research question by creating a game situation or a model of system conditions and then programming a computer to work through the moves and the interactions over and over again so that a statistically significant sample results. In Lovelock's Daisyworld, for example, there is no single equation by which one could predict the effect of black and white daisies through time on the temperature of a hypothetical planet. Rather, one must develop a computer program that contains equations specifying characteristics of daisy growth, effect on planetary albedo, increase in solar luminosity, and so forth, and then simply let it run.

Similarly, Axelrod reached his astonishing conclusions about the evolution of cooperation by pitting various behavioral strategies against one another in a round-robin computer tournament. The excerpts here show the results, drawn from The Evolution of Cooperation *by Robert Axelrod (copyright 1984 by Robert Axelrod and reprinted here by permission of the author and Basic Books, Inc., Publishers, New York).*

Advantages of Cooperation

Robert Axelrod

Under what conditions will cooperation emerge in a world of egoists without central authority? This question has intrigued people for a long time. And for good reason. We all know that people are not angels, and that they tend to look after themselves and their own first. Yet we also know that cooperation does occur and that our civilization is based upon it. But, in situations where each individual has an incentive to be selfish, how can cooperation ever develop? The answer each of us gives to this question has a fundamental effect on how we think and act in our social, political, and economic relations with others. And the answers that others give have a great effect on how ready they will be to cooperate with us.

The most famous answer was given over three hundred years ago by Thomas Hobbes *(in his book* Leviathan*)*. It was pessimistic. He argued that before governments existed, the state of nature was dominated by the problem of selfish individuals who competed on such ruthless terms that life was "solitary, poor, nasty, brutish, and short." In his view, cooperation could not develop without a central authority, and consequently a strong government was necessary. Ever since, arguments about the proper scope of government have often focused on whether one could, or could not, expect cooperation to emerge in a particular domain if there were not an authority to police the situation.

Today nations interact without central authority. Therefore the requirements for the emergence of cooperation have relevance to many of the central issues of international politics. A good example of the fundamental problem of cooperation is the case where two industrial nations have erected trade barriers to each other's exports. Because of the mutual advantages of free trade, both countries would be better off if these barriers were eliminated. But if either country were to unilaterally eliminate its barriers, it would find itself facing terms of trade that hurt its own economy. In fact, whatever one country does, the other country is better off retaining its own trade barriers. Therefore, the problem is that each country has an incentive to retain trade barriers, leading to a worse outcome than would have been possible had both countries cooperated with each other.

In everyday life, we may ask ourselves how many times we will invite acquaintances for dinner if they never invite us over in return. An executive in an organization does favors for another executive in order

Butchering at Gambell (Rie Muñoz, 1974)

to get favors in exchange. A journalist who has received a leaked news story gives favorable coverage to the source in the hope that further leaks will be forthcoming. A business firm in an industry with only one other major company charges high prices with the expectation that the other firm will also maintain high prices—to their mutual advantage and at the expense of the consumer.

The approach of this book is to investigate how individuals pursuing their own interests will act, followed by an analysis of what effects this will have for the system as a whole. The object of the enterprise is to develop a theory of cooperation that can be used to discover what is necessary for cooperation to emerge. By understanding the conditions that allow it to emerge, appropriate actions can be taken to foster the development of cooperation in a specific setting.

The basic problem occurs when the pursuit of self-interest by each leads to a poor outcome for all. To make headway in understanding the vast array of specific situations which have this property, a way is needed to represent what is common to these situations without becoming bogged down in the details unique to each. Fortunately, there is such a representation available: the famous Prisoner's Dilemma game.

The original story is that two accomplices to a crime are arrested and questioned separately. Either can defect against the other by confessing and hoping for a lighter sentence. But if both confess, their confessions are not as valuable. On the other hand, if both cooperate with each other by refusing to confess, the district attorney can only convict them on a minor charge. Assuming that neither player has moral qualms about, or fear of, squealing, the payoffs can form a Prisoner's Dilemma.

At this point let's take a break from Axelrod's book and move to an essay by Douglas Hofstadter from Metamagical Themas: Questing for the Essence of Mind and Pattern *(copyright 1985 by Basic Books, Inc). The book is a compilation of Hofstadter's writings as a* Scientific American *columnist in the early 1980s. (Excerpts here are reprinted by permission of the author and Basic Books, Inc., Publishers, New York.) Hofstadter, professor of cognitive science and computer science at Indiana University, offers a clear and witty description of the Prisoner's Dilemma. Here you will see the kind of writing that earned him a Pulitzer prize for his first book,* Gödel, Escher, Bach: an Eternal Golden Braid.

Rules of the Game

Douglas Hofstadter

Elegant mathematical structures can be as central to a serious modern worldview as are social concerns, and can deeply influence one's ways of thinking about anything—even such somber and colossal things as total nuclear obliteration. In order to comprehend that which is incomprehensible because it is too huge or too complex, one needs simpler models. Often, mathematics can provide the right starting point, which is why beautiful mathematical concepts are so pervasive in explanations of the phenomena of nature on the micro-level. They are now proving to be of great help also on a larger scale, as Robert Axelrod's lovely work on the Prisoner's Dilemma so impeccably demonstrates.

The Prisoner's Dilemma is poised about halfway between the Cube and Armageddon, in terms of complexity, abstraction, size, and seriousness. I submit that abstractions of this sort are direly needed in our times, because many people—even remarkably smart people—turn off when faced with issues that are too big. We need to make such issues graspable. To make them graspable and fascinating as well, we need to entice people with the beauties of clarity, simplicity, precision, elegance, balance, symmetry, and so on.

Those artistic qualities, so central to good science as well as to good insights about life, are the things that I have tried to explore and even to celebrate in *Metamagical Themas*. I hope that *Metamagical Themas* will help people to bring more clarity, precision, and elegance to their thinking about situations large and small. I also hope that it will inspire people to dedicate more of their energies to global problems in this lunatic but lovable world. . .

Life is filled with paradoxes and dilemmas. Sometimes it even feels as if the essence of living is the sensing—indeed the savoring—of paradox. Although all paradoxes seem somehow related, some paradoxes seem abstract and philosophical, while others touch life very directly. A very lifelike paradox is the so-called "Prisoner's Dilemma," discovered in 1950 by Melvin Dresher and Merrill Flood of the RAND Corporation. Albert W. Tucker wrote the first article on it, and in that article he gave it its now-famous name. I shall here present the Prisoner's Dilemma—first as a metaphor, then as a formal problem.

The original formulation in terms of prisoners is a little less clear to the uninitiated, in my experience, than the following one. Assume you possess copious quantities of some item (money, for example) and wish to obtain some amount of another item (perhaps stamps, groceries, diamonds). You arrange a mutually agreeable trade with the only dealer of that item known to you. You are both satisfied with the amounts you will be giving and getting. For some reason, though, your trade must take place in secret. Each of you agrees to leave a bag at a designated place in the forest, and to pick up the other's bag at the other's designated place. Suppose it is clear to both of you that the two of you will never meet or have further dealings with each other again.

Clearly, there is something for each of you to fear: namely, that the other one will leave an empty bag. Obviously if you both leave full bags, you will both be satisfied; but equally obviously, getting something for nothing is even more satisfying. So you are tempted to leave an empty bag. In fact, you can even reason it through quite rigorously this way: "If the dealer brings a full bag, I'll be better off having left an empty bag, because I'll have gotten all that I wanted and given away nothing. If the dealer brings an empty bag, I'll be better off having left an empty bag, because I'll not have been cheated. I'll have gained nothing but lost nothing either. Thus is seems that no matter what the dealer chooses to do, I'm better off leaving an empty bag. So I'll leave an empty bag."

The dealer, meanwhile, being in more or less the same boat (though at the other end of it), thinks analogous thoughts and comes to the parallel conclusion that it is best to leave an empty bag. And so both of you, with your impeccable (or impeccable-seeming) logic, leave empty bags, and go away empty-handed. How sad, for if you had both just cooperated, you could have each gained something you wanted to have. Does logic prevent cooperation? This is the issue of the Prisoner's Dilemma.

Let us now go back to the original metaphor and slightly alter its conditions. Suppose that both you and your partner very much want to have a regular supply of what the other has to offer, and so, before conducting your first exchange, you agree to carry on a lifelong exchange, once a month. You still expect never to meet face to face. In fact, neither of you has any idea how old the other one is, so you can't be very sure of how long this lifelong agreement may go on, but it seems safe to assume it'll go on for a few months anyway, and very likely for years.

Now, what do you do on your first exchange? Taking an empty bag seems fairly nasty as the opening of a relationship—hardly an effective way to build up trust. So suppose you take a full bag, and the dealer brings one as well. Bliss—for a month. Then you both must go back. Empty or full? Each month, you have to decide whether to "defect" (take an empty bag) or to "cooperate" (take a full one). Suppose that one month, unexpectedly, your dealer defects. Now what do you do? Will you suddenly decide that the dealer can never be trusted again, and from now on always bring empty bags, in effect totally giving up on the whole project forever? Or will you pretend you didn't notice, and continue being friendly? Or—will you try to punish the dealer by some number of defections of your own? One? Two? A random number? An increasing number, depending on how many defections you have experienced? Just how mad will you get?

This is the so-called iterated Prisoner's Dilemma. It is a very difficult problem. It can be, and has been, rendered more quantitative and in that form studied with the methods of game theory and computer simulation. How does one quantify it? One builds a "payoff matrix" presenting point values for the various alternatives. A typical one is shown here.

In this matrix, mutual cooperation earns both parties 2 points (the subjective value of receiving a full bag of what you need while giving up a full bag of what you have). Mutual defection earns you both zero points (the subjective value of gaining nothing and losing nothing, aside from making a vain trip out to the forest that month). Cooperating while the other defects stings: you get –1 point while the rat gets 4 points! Why so many? Because it is so pleasurable to get something for nothing. And of course, should you happen to be a rat some month when the dealer has cooperated, then you get 4 points and the dealer loses 1.

		Dealer	
		Cooperates	Defects
You	Cooperate	(2, 2)	(–1, 4)
	Defect	(4, –1)	(0, 0)

It is obvious that in a collective sense, it would be best for both of you to always cooperate. But suppose you have no regard whatsoever for the other. There is no "collective good" you are both working for. You are both supreme egoists. This means you have no feeling of friendliness or goodwill or compassion for the other player; you have no conscience; all you care about is amassing points, more and more and more of them.

This whole situation is highly relevant to questions in evolutionary biology. Can totally selfish and unconscious organisms living in a common environment come to evolve reliable cooperative strategies? Can cooperation emerge in a world of pure egoists? In a nutshell, can cooperation evolve out of noncooperation? If so, this has revolutionary import for the theory of evolution, for many of its critics have claimed that this was one place that it was hopelessly snagged.

With the rules of the game (and the dilemma itself) so clearly expounded, let's return to Robert Axelrod and the computer tournament he conducted for finding a winning strategy.

Staging a Computer Tournament

Robert Axelrod

Since the Prisoner's Dilemma is so common in everything from personal relations to international relations, it would be useful to know how best to act when in this type of setting. What is best depends in part on what the other player is likely to be doing. Further, what the other is likely to be doing may well depend on what the player expects you to do. To get out of this tangle, help can be sought by combing the research already done concerning the Prisoner's Dilemma for useful advice. Fortunately, a great deal of research has been done in this area.

Psychologists using experimental subjects have found that, in the iterated Prisoner's Dilemma, the amount of cooperation attained—and the specific pattern for attaining it—depend on a wide variety of factors relating to the context of the game, the attributes of the individual players, and relationship between the players. Since behavior in the game reflects so many important factors about people, it has become a standard way to explore questions in social psychology, from the effects of westernization in Central Africa, to the existence (or nonexistence) of aggression in career-oriented women, and to the differential consequences of abstract versus concrete thinking styles. In the last fifteen years, there have been hundreds of articles on the Prisoner's Dilemma cited in *Psychological Abstracts*. The iterated Prisoner's Dilemma has become the *E. coli* of social psychology.

Unfortunately, none of the literature on the Prisoner's Dilemma reveals very much about how to play the game well. To learn more about how to choose effectively in an iterated Prisoner's Dilemma, a new approach is needed. Such an approach would have to draw on people who have a rich understanding of the strategic possibilities inherent in a non-zero-sum setting, a situation in which the interests of

		Column Player	
		Cooperate	Defect
Row Player	Cooperate	$R = 3, R = 3$ Reward for mutual cooperation	$S = 0, T = 5$ Sucker's payoff, and temptation to defect
	Defect	$T = 5, S = 0$ Temptation to defect and sucker's payoff	$P = 1, P = 1$ Punishment for mutual defection

Note: The payoffs to the row chooser are listed first.

the participants partially coincide and partially conflict. A computer tournament for the study of effective choice in the iterated Prisoner's Dilemma meets these needs. In a computer tournament, each entrant writes a program that embodies a rule to select the cooperative or noncooperative choice on each move. The program has available to it the history of the game so far, and may use this history in making a choice.

Wanting to find out what would happen, I invited professional game theorists to send in entries to just such a computer tournament. Most of the entrants were recruited from those who had published articles on game theory in general or the Prisoner's Dilemma in particular. The fourteen submitted entries came from five disciplines: psychology, economics, political science, mathematics, and sociology.

The tournament was structured as a round robin, meaning that each entry was paired with each other entry. As announced in the rules, each entry was also paired with its own twin and with "Random," a program that randomly cooperates and defects with equal probability. Each game consisted of exactly two hundred moves. The payoff matrix shown here awarded both players 3 points for mutual cooperation, and 1 point for mutual defection. If one player defected while the other player cooperated, the defecting player received 5 points and the cooperating player received 0 points.

The Winner: Tit-for-Tat

"Tit-for-Tat," submitted by Professor Anatol Rapoport of the University of Toronto, won the tournament. This was the simplest of all submitted programs and it turned out to be the best! Tit-for-Tat starts with a cooperative choice, and thereafter does what the other player did on the previous move. This decision rule is probably the most widely known and most discussed rule for playing the Prisoner's Dilemma. It is easily understood and easily programmed. It is known to elicit a good degree of cooperation when played with humans. As an entry in a computer tournament, it has the desirable properties that it is not very exploitable and that it does well with its own twin.

Not surprisingly, many of the entrants used the Tit-for-Tat principle and tried to improve upon it. The striking fact is that none of the more complex programs submitted was able to perform as well as the original, simple Tit-for-Tat. Analysis of the results showed that neither the discipline of the author, the brevity of the program—nor its length—accounts for a rule's relative success. What does?

Surprisingly, there is a single property which distinguishes the relatively high-scoring entries from the relatively low-scoring entries. This is the property of being "nice," which is to say never being the first to defect. The nice rules did well in the tournament largely because they did so well with each other, and because there were enough of them to raise substantially each other's average score. As long as the other player did not defect, each of the nice rules was certain to continue cooperating until virtually the end of the game. But what happened if there was a defection? Different rules responded quite differently, and their response was important in determining their overall success. A key concept in this regard is the "forgiveness" of a decision rule. "Forgiveness" of a rule can be informally described as its propensity to cooperate in the moves after the other player has defected.

Of all the nice rules, the one that scored lowest was also the one that was least forgiving. This is "Friedman," a totally unforgiving rule that employs permanent retaliation. It is never the first to defect, but once the other defects even once, Friedman defects from then on. In contrast, the winner, Tit-for-Tat, is unforgiving for one move, but thereafter is totally forgiving of that defection. After one punishment, it lets bygones be bygones.

One of the main reasons why the rules that are not nice did not do well in the tournament is that most of the rules in the tournament were not very forgiving. Consider the case of "Joss," a sneaky rule that tries to get away with an occasional defection. This decision rule is a variation of Tit-for-Tat. Like Tit-for-Tat, it always defects immediately after the other player defects. But instead of always cooperating after the other player cooperates, 10 percent of the time it defects after the other player cooperates. Thus it tries to sneak in an occasional exploitation of the other player. However, if both sides retaliate in the way that Joss and Tit-for-Tat do, it does not pay to be as greedy as Joss.

Despite the fact that none of the attempts at more or less sophisticated decision rules was an improvement on Tit-for-Tat, it was easy to find several rules that would have performed substantially better than Tit-for-Tat in the environment of that particular tournament. The existence of these rules should serve as a warning against the facile belief that an eye for an eye is necessarily the best strategy. There are at least three rules that would have won the tournament if submitted. The sample program sent to prospective contestants to show them how to make a submission would in fact have won the tournament if anyone had simply clipped it and mailed it in! But no one did. The sample program defects only if the other player defected on the previous two moves. It is a more forgiving version of Tit-for-Tat in that it does not punish isolated defections. The excellent performance of this Tit-for-Two-Tats rule highlights the fact that a common error of the contestants was to expect that gains could be made from being relatively less forgiving than Tit-for-Tat, whereas in fact there were big gains to be made from being even more forgiving.

Round Two

The analysis of the tournament results indicate that there is a lot to be learned about coping in an environment of mutual power. Even expert strategists from political science, sociology, economics, psychology, and mathematics made the systematic errors of being too competitive for their own good, not being forgiving enough, and being too pessimistic about the responsiveness of the other side. The effectiveness of a particular strategy depends not only on its own characteristics, but also on the nature of the other strategies with which it must interact. For this reason, the results of a single tournament are not definitive. Therefore, a second round of the tournament was conducted.

The second round presumably began at a much higher level of sophistication than the first round, and its results could be expected to be that much more valuable as a guide to effective choice in the Prisoner's Dilemma. The second round was also a dramatic improvement over the first round in sheer size of the tournament. The response was far greater than anticipated. There was a total of sixty-two entries from six countries. The contestants ranged from a ten-year-old computer hobbyist to professors of computer science, physics, economics, psychology, mathematics, sociology, political science, and evolutionary biology. The countries represented were the United States, Canada, Great Britain, Norway, Switzerland, and New Zealand.

Tit-for-Tat was the simplest program submitted in the first round, and it won the first round. It was the simplest submission in the second round, and it won the second round. Even though all the entrants to the second round knew that Tit-for-Tat had won the first round, no one was able to design an entry that did any better. And even though an explicit tournament rule allowed anyone to submit any program, even one authored by someone else, only one person submitted Tit-for-Tat. This was Anatol Rapoport, who submitted it the first time. *(Douglas Hofstadter describes Rapoport as "a psychologist and philosopher." Robert Wright in his book* Three Scientists and Their Gods *reports that*

Rapoport worked as a mathematical biologist early in his career, and that he later became known for his work in game theory. Rapoport is also recognized as one of the initial founders, along with Ludwig von Bertalanffy and Kenneth Boulding, of the Society for General Systems Research, which perhaps explains why he is difficult to peg.)

Once again, none of the personal attributes of the contestants correlated significantly with the performance of the rules. The professors did not do significantly better than the others, nor did the Americans. Those who wrote in "Fortran" rather than "Basic" did not do significantly better either, even though the use of Fortran would usually indicate access to something more than a bottom-of-the-line microcomputer.

In the second round, there was once again a substantial correlation between whether a rule was nice and how well it did. Of the top fifteen rules, all but one were nice (and that one ranked eighth). Of the bottom fifteen rules, all but one were not nice. The overall correlation between whether a rule was nice and its tournament score was a substantial .58.

A property that distinguishes well among the nice rules themselves is how promptly and how reliably they responded to a challenge by the other player. A rule can be called "retaliatory" if it immediately defects after an "uncalled for" defection from the other. Exactly what is meant by "uncalled for" is not precisely determined. The point, however, is that unless a strategy is incited to an immediate response by a challenge from the other player, the other player may simply take more and more frequent advantage of such an easygoing strategy.

There were a number of rules in the second round of the tournament that deliberately used controlled numbers of defections to see what they could get away with. To a large extent, what determined the actual rankings of the nice rules was how well they were able to cope with these challengers. Tit-for-Tat combines these desirable properties. It is nice, forgiving, and retaliatory. It is never the first to defect; it forgives an isolated defection after a single response; but it is always incited by a defection no matter how good the interaction has been so far. Tit-for-Two-Tats was submitted by an evolutionary biologist from the United Kingdom, John Maynard Smith. But it came in only twenty-fourth.

Lessons of Tit-for-Tat

What can be said for the empirical successes of Tit-for-Tat is that it is a very robust rule: it does very well over a wide range of environments. Part of its success might be that other rules anticipate its presence and are designed to do well with it. Doing well with Tit-for-Tat requires cooperating with it, and this in turn helps Tit-for-Tat. Any rule which tries to take advantage of Tit-for-Tat will simply hurt itself.

What accounts for Tit-for-Tat's robust success is its combination of being nice, retaliatory, forgiving, and clear. Its niceness prevents it from getting into unnecessary trouble. Its retaliation discourages the other side from persisting whenever defection is tried. Its forgiveness helps restore mutual cooperation. And its clarity makes it intelligible to the other player, thereby eliciting long-term cooperation.

The analysis of the Computer Tournament and the results of the theoretical investigations provide some useful information about what strategies are likely to work under different conditions and why. The purpose of this section is to translate these findings into advice for a player. The advice takes the form of four simple suggestions for how to do well in a durable iterated Prisoner's Dilemma:

- Don't be envious.
- Don't be the first to defect.
- Reciprocate both cooperation and defection.
- Don't be too clever

"Don't be envious"

People are used to thinking about zero-sum interactions. In these settings, whatever one person wins, another loses. A good example is a chess tournament. In order to do well, the contestant must do better than the other player in the game most of the time. A win for White is necessarily a loss for Black. But most of life is not zero-sum. Generally, both sides can do well, or both can do poorly. Mutual cooperation is often possible, but not always achieved. That is why the Prisoner's Dilemma is such a useful model for a wide variety of everyday situations.

In my classes, I have often had pairs of students play the Prisoner's Dilemma for several dozen moves. I tell them that the object is to score well for themselves, as if they were getting a dollar a point. I also tell them that it should not matter to them whether they score a little better or a little worse than the other player, so long as they can collect as many "dollars" for themselves as possible.

These instructions simply do not work. The students look for a standard of comparison to see if they are doing well or poorly. People tend to resort to the standard of comparison that they have available—and this standard is often the success of the other player relative to their own success. This standard leads to envy. And envy leads to attempts to rectify any advantage the other player has attained. In this form of Prisoner's Dilemma, rectification of the other's advantage can only be done by defection. But defection leads to more defection and to mutual punishment. So envy is self-destructive.

A better standard of comparison is how well you are doing relative to how well someone else could be doing in your shoes. Given the strategy of the other player, are you doing as well as possible? Could someone else in your situation have done better with this other player? This is the proper test of successful performance.

The Cardsharps (M. M. da Caravaggio, 1594)

Tit-for-Tat won the tournament because it did well in its interactions with a wide variety of other strategies. Yet Tit-for-Tat never once scored better in a game than the other player! In fact, it can't. It lets the other player defect first, and it never defects more times than the other player has defected. Therefore, Tit-for-Tat achieves either the same score as the other player, or a little less. Tit-for-Tat won the tournament, not by beating the other player, but by eliciting behavior from the other player which allowed both to do well. Tit-for-Tat was so consistent at eliciting mutually rewarding outcomes that it attained a higher overall score than any other strategy.

"Be nice"

Both the tournament and the theoretical results show that it pays to cooperate as long as the other player is cooperating. The single best predictor of how well a rule performed was whether or not it was nice, which is to say, whether or not it would ever be the first to defect. In the first round, each of the top eight rules were nice, and not one of the bottom seven were nice. In the second round, all but one of the top fifteen rules were nice (and that one ranked eighth). Of the bottom fifteen rules, all but one were not nice.

"Practice reciprocity"

The extraordinary success of Tit-for-Tat leads to some simple, but powerful advice: practice reciprocity. After cooperating on the first move, Tit-for-Tat simply reciprocates whatever the other player did on the previous move. The beauty of the reciprocity of Tit-for-Tat is that it is good in such a wide range of circumstances. In fact, Tit-for-Tat is very good at discriminating between rules which will return its own initial cooperation and those which will not. This allows it to invade a world of meanies in the smallest possible cluster. Moreover, it will reciprocate a defection as well as cooperation, making it provocable. Being provocable is actually required for a nice rule like Tit-for-Tat to resist invasion.

In responding to a defection from the other player, Tit-for-Tat represents a balance between punishing and being forgiving. This suggests the question of whether always doing exactly one-for-one is the most effective balance. It is hard to say because rules with slightly different balances were not submitted. What is clear is that extracting more than one defection for each defection of the other risks escalation. On the other hand, extracting less than one-for-one risks exploitation.

Tit-for-Two-Tats is the rule that defects only if the other player has defected in both of the previous two moves. Therefore it returns one-for-two. This relatively forgiving rule would have won the first round of the Computer Tournament had it been submitted. Yet in the second round, when Tit-for-Two-Tats was actually submitted, it did not even score in the top third. The reason is that the second round contained some rules that were able to exploit its willingness to forgive isolated defections.

The moral of the story is that the precise level of forgiveness that is optimal depends upon the environment. In particular, if the main danger is unending mutual recriminations, then a generous level of forgiveness is appropriate. But, if the main danger is from strategies that are good at exploiting easygoing rules, then an excess of forgiveness is costly. While the exact balance will be hard to determine in a given environment, the evidence of the tournament suggests that something approaching a one-for-one response to defection is likely to be quite effective in a wide range of settings. Therefore it is good advice to a player to reciprocate defection as well as cooperation.

"Don't be too clever"

The tournament results show that in a Prisoner's Dilemma situation it is easy to be too clever. The very sophisticated rules did not do better than the simple ones. In fact, the so-called maximizing rules often did poorly because they got into a rut of mutual defection. A common problem with these rules is that they used complex methods of making inferences about the other player—and these inferences were wrong. Part of the problem was that a trial defection by the other player was often taken to imply that the other player could not be enticed into cooperation. But the heart of the problem was that these maximizing rules did not take into account that their own behavior would lead the other player to change. *(This problem is called feedback. It is one of the key features of "organized complex systems" identified by Ludwig von Bertalanffy, the founder of general systems theory. Feedback between living organisms and the earth's climate and chemistry is a feature that James Lovelock attributes to Gaia.)*

In deciding whether to carry an umbrella, we do not have to worry that the clouds will take our behavior into account. We can do a calculation about the chance of rain based on past experience. Likewise in a zero-sum game, such as chess, we can safely use the assumption that the other player will pick the most dangerous move that can be found, and we can act accordingly. Therefore it pays for us to be as sophisticated and as complex in our analysis as we can.

Non-zero-sum games, such as the Prisoner's Dilemma, are not like this. Unlike the clouds, the other player can respond to your own choices. And unlike the chess opponent, the other player in a Prisoner's Dilemma should not be regarded as someone who is out to defeat you. The other player will be watching your behavior for signs of whether you will reciprocate cooperation or not, and therefore your own behavior is likely to be echoed back to you.

Rules that try to maximize their own score while treating the other player as a fixed part of the environment ignore this aspect of interaction, no matter how clever they are in calculating under their limiting assumptions. Therefore, it does not pay to be clever in modeling the other player if you leave out the reverberating process in which the other player is adapting to you, you are adapting to the other, and then the other is adapting to your adaptation and so on. This is a difficult road to follow with much hope for success. Certainly none of the more or less complex rules submitted in either round of the tournament was very good at it.

Another way of being too clever is to use a strategy of permanent retaliation. Permanent retaliation may seem clever because it provides the maximum incentive to avoid defection. But it is too harsh for its own good.

There is yet a third way in which some of the tournament rules are too clever: they employ a probabilistic strategy that is so complex that it cannot be distinguished by the other strategies from a purely random choice. In other words, too much complexity can appear to be total chaos. If you are using a strategy which appears random, then you also appear unresponsive to the other player. If you are unresponsive, then the other player has no incentive to cooperate with you. So being so complex as to be incomprehensible is very dangerous.

One way to account for Tit-for-Tat's great success in the tournament is that it has great clarity: it is eminently comprehensible to the other player. When you are using Tit-for-Tat, the other player has an excellent chance of understanding what you are doing. Your one-for-one response to any defection is an easy pattern to appreciate. Your future behavior can then be predicted. Once this happens, the other player can easily see that the best way to deal with Tit-for-Tat is to cooperate with it.

Once again, there is an important contrast between a zero-sum game like chess and a non-zero-sum game like the iterated Prisoner's Dilemma. In chess, it is useful to keep the other player guessing about your intentions. The more the other player is in doubt, the less efficient will be his or her strategy. Keeping one's intentions hidden is useful in a zero-sum setting where any inefficiency in the other player's behavior will be to your benefit. But in a non-zero-sum setting it does not always pay to be so clever. In the iterated Prisoner's Dilemma, you benefit from the other player's cooperation. The trick is to encourage that cooperation. A good way to do it is to make it clear that you will reciprocate. Words can help here, but as everyone knows, actions speak louder than words.

10 Game Theory and Biology

"After thinking about the evolution of cooperation in a social context, I realized that the finding also had implications for biological evolution. So I collaborated with a biologist, William Hamilton, to develop the biological implications of these strategic ideas. This resulted in a paper published in Science *(1981). The paper has been awarded the Newcomb Cleveland Prize of the American Association for the Advancement of Science." So writes Robert Axelrod in* The Evolution of Cooperation *(1984). Axelrod and Hamilton modified their scientific paper for presentation as a chapter in Axelrod's book, excerpted below. Extensive citations in the original text have been eliminated in these excerpts.*

Cooperation in Nature

Robert Axelrod with William D. Hamilton

The theory of biological evolution is based on the struggle for life and the survival of the fittest. Yet cooperation is common between members of the same species and even between members of different species. To account for the manifest existence of cooperation and related group behavior, such as altruism and restraint in competition, evolutionary theory has recently acquired two kinds of extension. These extensions are, broadly, genetical kinship theory and reciprocity theory.

Most of the recent activity, both in fieldwork and in further developments of theory, has been on the side of kinship. Formal approaches have varied, but kinship theory has increasingly taken a gene's-eye view of natural selection. A gene, in effect, looks beyond its mortal bearer to the potentially immortal set of its replicas existing in other related individuals. If the players are sufficiently closely related, altruism can benefit reproduction of the set, despite losses to the individual altruist. In accord with this theory's predictions, almost all clear cases of altruism, and most observed cooperation—apart from their appearance in the humans species—occur in contexts of high relatedness, usually between immediate family members. The evolution of the suicidal barbed sting of the honeybee worker could be taken as a paradigm for this line of theory.

Conspicuous examples of cooperation (although almost never of ultimate self-sacrifice) also occur where relatedness is low or absent. Mutually advantageous symbioses offer striking examples such as these: the fungus and alga that compose a lichen; the ants and ant-acacias, where the trees house and feed the ants which, in turn, protect the trees; and the fig wasps and fig tree, where wasps, which are parasites of fig flowers, serve as the tree's sole means of pollination and seed set. Although kinship may be involved, symbioses mainly illustrate the other recent extension of evolution theory—the theory of reciprocity.

Cooperation itself has received comparatively little attention from biologists since the 1971 pioneer account of Robert L. Trivers ("The Evolution of Reciprocal Altruism"). But an associated issue, concerning restraint in conflict situations, has been developed theoretically. In this connection, a new concept—that of an "evolutionarily stable strategy" (ESS)—has been formally developed by John Maynard Smith and others. *(A good introduction to this field is Maynard Smith's 1982 book,* Evolution and the Theory of Games.*)* Cooperation in the more normal sense has remained clouded by certain difficulties, particularly those concerning the initiation of cooperation from a previously asocial state and its stable maintenance once established. A formal theory of cooperation is increasingly needed.

Strategies for the Brainy and the Brainless

Surprisingly, there is a broad range of biological reality that is encompassed by this game-theoretic approach. To start with, an organism does not need a brain to employ a strategy. Bacteria, for example, have a basic capacity to play games in that (1) bacteria are highly responsive to selected aspects of their environment, especially their chemical environment; (2) this implies that they can respond differentially to what other organisms around them are doing; (3) these conditional strategies of behavior can certainly be inherited; and (4) the behavior of a bacterium can affect the fitness of other organisms around it, just as the behavior of other organisms can affect the fitness of a bacterium.

While the strategies can easily include differential responsiveness to recent changes in the environment or to cumulative averages over time, in other ways their range of responsiveness is limited. Bacteria cannot "remember" or "interpret" a complex past sequence of changes, and they probably cannot distinguish alternative origins of adverse or beneficial changes. Some bacteria, for example, produce their own antibiotics, called bacteriocins. These are harmless to bacteria of the producing strain, but are destructive to others. A bacterium might easily have production of its own bacteriocin dependent on the perceived presence of like hostile products in its environment, but it could not aim the toxin produced toward an offending initiator.

As one moves up the evolutionary ladder in neural complexity, game-playing behavior becomes richer. The intelligence of primates, including humans, allows a number of relevant improvements: a more complex memory, more complex processing of information to determine the next action as a function of the interaction so far, a better estimate of the probability of future interaction with the same individual, and a better ability to distinguish between different individuals. The discrimination of others may be among the most important of abilities because it allows one to handle interactions with many individuals without having to treat them all the same, thus making possible the rewarding of cooperation from one individual and the punishing of defection from another.

Judging Success

Turning to the development of the theory, the evolution of cooperation can be conceptualized in terms of three separate questions:

Robustness. What type of strategy can thrive in a variegated environment composed of others using a wide variety of more or less sophisticated strategies?

Stability. Under what conditions can such a strategy, once fully established, resist invasion by mutant strategies?

Initial viability. Even if a strategy is robust and stable, how can it ever get a foothold in an environment which is predominantly noncooperative?

The computer tournament showed that Tit-for-Tat's strategy of cooperation based on reciprocity was extremely robust. An ecological analysis found that as less successful rules were displaced, Tit-for-Tat continued to do well with the rules which initially did well. Thus cooperation based on reciprocity can thrive in a variegated environment. Once a strategy has been adopted by the entire population, the question of evolutionary stability deals with whether it can resist invasion by a mutant strategy. Tit-for-Tat is in fact evolutionarily stable if and only if the interactions between the individuals have a sufficiently large probability of continuing.

Tit-for-Tat is not the only strategy that can be evolutionarily stable. In fact, "Always-Defect" is evolutionarily stable no matter what the probability is of interaction continuing. This raises the problem of how an evolutionary trend to cooperative behavior could ever have started in the first place. Genetic kinship theory suggests a plausible escape from the equilibrium of Always-Defect. Close relatedness of players permits true altruism—sacrifice of fitness by one individual for the benefit of another. True altruism can evolve when the conditions of cost, benefit, and relatedness yield net gains for the altruism-causing genes that are resident in the related individuals.

When the probability of two individuals meeting each other again is sufficiently high, cooperation based on reciprocity can thrive and be evolutionarily stable in a population with no relatedness at all. A case of cooperation that fits this scenario, at least on first evidence, has been

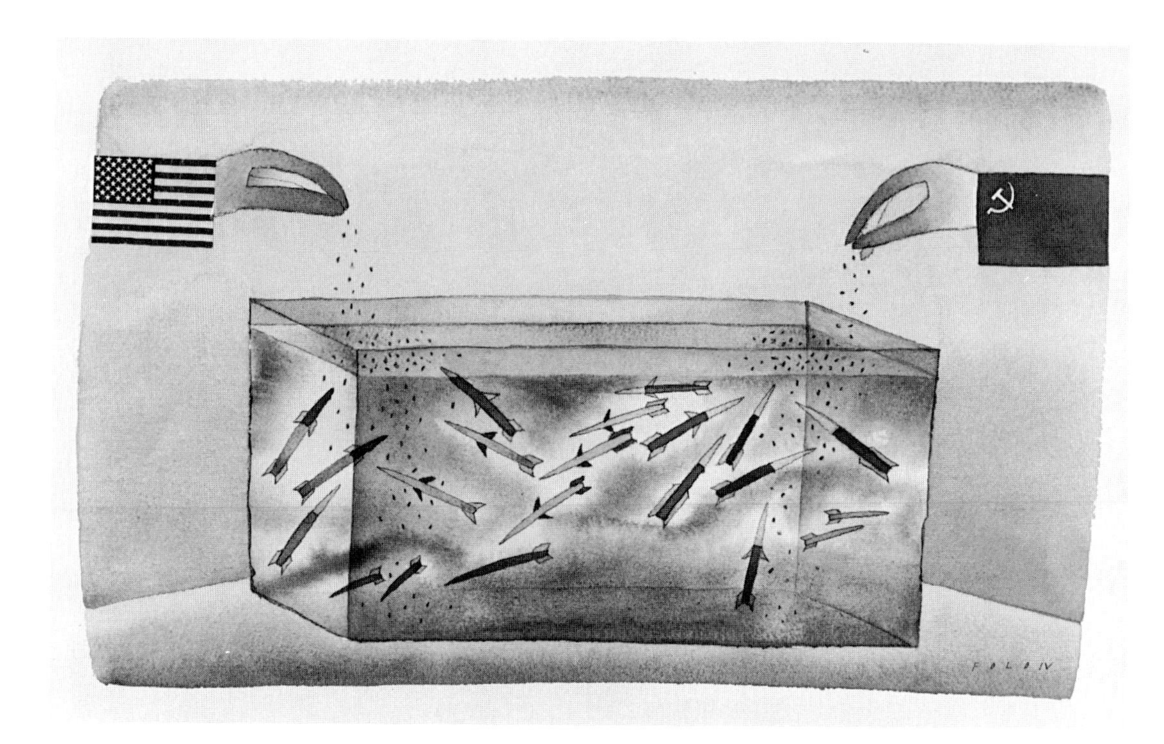

Always More (Jean-Michel Folon, 1983)

discovered in the spawning relationships in a sea bass. These fish have the sexual organs of both the male and the female. They form pairs and roughly may be said to take turns at being the high investment partner (laying eggs) and low investment partner (providing sperm to fertilize eggs). Up to ten spawnings occur in a day and only a few eggs are provided each time. Pairs tend to break up if sex roles are not divided evenly.

An individual must not be able to get away with defecting without the other individuals being able to retaliate effectively. The response requires that the defecting individual not be lost in a sea of anonymous others. Higher organisms avoid this problem by their well-developed ability to recognize many different individuals of their species, but lower organisms must rely on mechanisms that drastically limit the number of different individuals or colonies with which they can interact effectively.

Recognizing Friend or Foe

When an organism is not able to recognize the individual with which it had a prior interaction, a substitute mechanism is to make sure that all of its interactions are with the same player. This can be done by maintaining continuous contact with the other. This method is applied in most mutualisms, situations of close association of mutual benefit between members of different species. Examples include a hermit crab and its sea-anemone partner, a cicada and the varied colonies of microorganisms housed in its body, or a tree and its mycorrhizal fungi.

Another mechanism for avoiding the need for recognition is to guarantee the uniqueness of the pairing of players by employing a fixed place of meeting. Consider, for example, mutualisms based on cleaning in which a small fish or a crustacean removes and eats parasites from the body (or even from the inside of the mouth) of a larger fish that is its potential predator. These aquatic cleaner mutualisms occur in coastal and reef situations where animals live in fixed home ranges or territories. They seem to be unknown in the free-mixing circumstances of the open sea.

Other mutualisms are also characteristic of situations where continued association is likely, and normally they involve quasi-permanent pairing of individuals, or of inbred or asexual stocks, or of individuals with such stocks. Conversely, conditions of free-mixing, and transitory pairing conditions where recognition is impossible, are much more likely to result in exploitation—parasitism, disease, and the like. Thus, whereas ant colonies participate in many symbioses and are sometimes largely dependent on them, honeybee colonies—which are much less permanent in place of abode—have no known symbionts but many parasites.

The small freshwater animal *Chlorohydra viridissima* has a permanent, stable association with green algae that are always naturally found in its tissues and are very difficult to remove. In this species the alga is transmitted to new generations by way of the egg. *Hydra vulgaris* and *H. attentuata* also associate with algae but do not have egg transmission. In these species it is said that infection is preceded by enfeeblement of the animals and is accompanied by pathological symptoms indicating a definite parasitism by the plant. Again, it is seen that impermanence of association tends to destabilize symbiosis.

In species with a limited ability to discriminate between other members of the same species, reciprocal cooperation can be stable with the aid of a mechanism that reduces the amount of discrimination necessary. Territoriality can serve this purpose. In the case of male territorial birds, songs are used to allow neighbors to recognize each other. Consistent with the theory, such male territorial birds show much more aggressive reactions when the song of an unfamiliar male rather than a neighbor is reproduced nearby.

Reciprocal cooperation can be stable with a larger range of individuals if discrimination can cover a wide variety of others with less reliance on supplementary cues such as location. In humans this ability is well developed, and is largely based on the recognition of faces. The extent to which this function has become specialized is revealed by a brain disorder called prosopagnosia. A normal person can name someone from facial features alone, even if the features have changed substantially over the years. People with prosopagnosia are not able to make this association, but have few other neurological symptoms other than a loss of some part of the visual field. The lesions responsible for the disorder occur in an identifiable part of the brain: the underside of both occipital lobes, extending forward to the inner surface of the temporal lobes. This localization of cause, and specificity of effect, indicates that the recognition of individual faces has been an important enough task for a significant portion of the brain's resources to be devoted to it.

When Symbionts Become Parasites

Just as the ability to recognize the other player is invaluable in extending the range of stable cooperation, the ability to monitor cues for the likelihood of continued interaction is helpful as an indication of when reciprocal cooperation is or is not stable. In particular, when the relative importance of future interactions falls below the threshold for stability, it will no longer pay to reciprocate the other's cooperation. Illness in one partner leading to reduced viability would be one detectable sign of a declining probability. Both animals in a partnership would then be expected to become less cooperative. Aging of a partner would be very like disease in this respect, resulting in an incentive to defect so as to take a one-time gain when the probability of future interaction becomes small enough.

These mechanisms could operate even at the microbial level. Any symbiont that still has a chance to spread to other hosts by some process of infection would be expected to shift from mutualism to parasitism when the probability of continued interaction with the original host lessened. In the more parasitic phase, it could exploit the host more severely by producing more of the forms able to disperse and infect. This phase would be expected when the host is severely injured, has contracted some other wholly parasitic infection that threatens death, or when it manifests signs of age. In fact, bacteria that are normal and seemingly harmless or even beneficial in the gut can be found contributing to sepsis in the body when the gut is perforated, implying a severe wound. And normal inhabitants of the body surface (like *Candida albicans*) can become invasive and dangerous in either sick or elderly persons. It is possible also that this argument has some bearing on the causes of cancer, insofar as it turns out to be due to viruses potentially latent in the genome. Cancers do tend to have their onset at ages when the chances of transmission from one generation to the next are rapidly declining.

In conclusion, Darwin's emphasis on individual advantage has been formalized in terms of game theory. This formulation establishes conditions under which cooperation in biological systems based on reciprocity can evolve even without foresight by the participants.

Axelrod's book ranges far more widely than the excerpts presented here suggest. In one of the most fascinating chapters, Axelrod uses game theory to reveal why trench warfare during World War I included such uncombatlike behavior as "live and let live." Another chapter called "Advice for Participants and Reformers" shows how insights drawn from the computer tournament can be put to practical use in nurturing a more cooperative world. Axelrod concludes,

Robert Axelrod

The advice to players of the Prisoner's Dilemma might serve as good advice to national leaders as well: don't be envious, don't be the first to defect, reciprocate both cooperation and defection, and don't be too clever. Likewise, the techniques for promoting cooperation in the Prisoner's Dilemma might also be useful in promoting cooperation in international politics. The core of the problem of how to achieve rewards from cooperation is that trial and error in learning is slow and painful. The conditions may all be favorable for long-run developments, but we may not have the time to wait for blind processes to move us slowly toward mutually rewarding strategies based upon reciprocity. Perhaps if we understand the process better, we can use our foresight to speed up the evolution of cooperation.

V *Nature, Nurture, and Sociobiology*

A devil, a born devil,
On whose nature
Nurture can never stick!
On whom my pains,
Humanely taken, all, all lost, quite lost!

—Prospero on Caliban from Shakespeare's *Tempest*

To say we are evolved to serve the interests of our genes in no way suggests that we are obliged to serve them. Evolution is surely most deterministic for those still unaware of it.

—Richard D. Alexander

11 *From Ants to Anthropology: E. O. Wilson*

Game theory presents cooperation as a winning strategy shaped by the iron fist of natural selection. By cooperating with others, an organism can best promote its own survival. Now game theory has become an important tool for an altogether new field of biology: sociobiology.

Edward O. Wilson, eminent entomologist and winner of the National Medal of Science, is best known today as a champion of biodiversity, as codeveloper with Robert MacArthur of the theory of island biogeography (which is a powerful tool for predicting species extinctions due to habitat fragmentation), and as the originator of sociobiology. This chapter will focus on his contributions in the latter realm. As you will see, sociobiology is perhaps the most contentious field in all of biology—with colossal ramifications for how we humans view ourselves as social beings. Lewis Thomas argues that biology education ought to include the study of sociobiology. (The excerpt is from "Humanities and Science," copyright 1983 by Lewis Thomas in Late Night Thoughts on Listening to Mahler's Ninth Symphony, *reprinted here by permission of the author and Viking Penguin, a division of Penguin Books USA Inc.)*

Lewis Thomas

The running battle now in progress between the sociobiologists and the antisociobiologists is a marvel for students to behold, close up. To observe, in open-mouthed astonishment, the polarized extremes, one group of highly intelligent, beautifully trained, knowledgeable, and imaginative scientists maintaining that all sorts of behavior, animal and human, are governed exclusively by genes, and another group of equally talented scientists saying precisely the opposite and asserting that all behavior is set and determined by the environment, or by culture, and both sides brawling in the pages of periodicals such as *The New York Review of Books*, is an educational experience that no college student should be allowed to miss. The essential lesson to be learned has nothing to do with the relative validity of the facts underlying the argument, it is the argument itself that is the education: we do not yet know enough to settle such questions.

Before we examine arguments pro and con, an essay will introduce the scientist at the heart of the controversy. This biography of Edward O. Wilson is drawn from the 1988 book by Robert Wright, Three Scientists and Their Gods. *Wright is unusually well positioned to deal with the issue of sociobiology. He once specialized in science writing (which culminated in a 1986 National Magazine Award for his "The Information Age" column in* The Sciences*). He is now a senior editor at* The New Republic, *which offers readers incisive social and political commentary. Excerpts from* Three Scientists and Their Gods, *copyright 1988 by Robert Wright, are reprinted here by permission of the author and Times Books, a division of Random House, Inc.*

The Young Naturalist

Robert Wright

It was in boyhood that Wilson first experienced "the naturalist's trance," as he called it in his partly autobiographical book *Biophilia.* Alone in the woods, free from the demands of human society, he could enter a world oblivious to his. He still can. "I need a bigger fix now," he says. "When I was eighteen I could get all excited just getting out into a swamp in Alabama. Now it takes a rain forest in Brazil. But I still get the same rush, the same emotional high—hoping to find new things, exploring."

Occasionally Ed managed to turn a schoolmate into a "part-time zoologist." One convert was Ellis MacLeod, now a professor of entomology at the University of Illinois. In the fifth grade, MacLeod and Wilson lived blocks apart in Washington, D.C. On weekends they rode the bus to the Smithsonian Natural History Museum or walked to the National Zoo or to Rock Creek Park. At the park, wielding butterfly nets made of broomsticks, coat hangers, and cheesecloth, they bagged red admirals, fritillaries, and mourning cloaks, which would later be killed, pressed, and immortalized as representatives of their species.

Ellis remembers Ed writing stories whose heroes were animals and reading them before the entire student body. In spite of literary renown, though, Ed was shy. "I don't recall that he had many close friends, if any, and I didn't either," says MacLeod. "We were both a couple of creeps by contemporary standards—you know, those crazy sorts of people who go around picking up insects. I know Ed, in sixth grade recess, once just absolutely blew every kid's mind by letting a wolf spider walk over his hand." By junior high, when MacLeod visited the Wilsons in Alabama, Ed had moved up from spiders to "beetles, big showy things . . . praying mantises." A year or two later he assembled a backyard collection of snakes. They lived in chicken-wire cages and dined on the fish and tree frogs that Wilson dutifully procured. Inquiring neighbors were treated to tours of the collection. "I was rather locally famous for my devotion to snakes," Wilson recalls.

But his fate was not to study snakes. One day, at age seven, the year of his parents' divorce, he was fishing from a wharf and he yanked one of his catches out of the water with such uncontrolled force that its fin entered his right eye. The result was a traumatic cataract that left the eye nearly useless, permanently dulling his perception of depth. But the vision in his left eye remained acute, especially at close range, a fact that steered him toward the study of the small-scale. "I am the last to spot a hawk sitting in a tree," he has written, "but I can examine the hairs and contours of an insect's body without the aid of a magnifying glass."

Ants first caught Wilson's attention at Rock Creek Park. The memory is still clear: upon picking apart a rotting tree stump, he beheld a dense colony of *Acanthomyops*. Their golden bodies glistened in the sun, and the air was pungent with their characteristic smell, citronella. The next year, after moving to Alabama, he embarked on a systematic study of ants, collecting specimens of *Odontomachus insularis* and *Iridomyrmex humilis*. His career aspirations were beginning to take shape: he would work for the U.S. Department of Agriculture's extension service—drive around in one of those government-green pickup trucks and advise farmers about insect enemies and allies. At age thirteen Wilson made "my first publishable observation" (though he would not publish it for years)—that the Mobile Alabama area was rife with a fire ant once confined to South America: *Solenopsis invicta*.

The Premier Social Insects

"On several counts ants can be regarded as the premier social insects," Wilson wrote in *The Insect Societies*, the book that, upon publication in 1971, established him as one of the world's foremost entomologists. In elaboration, he speaks with awe about the ability of these simple little creatures to constitute societies of such complexity, size, and relentless efficiency. Consider *Eciton burchelli*. These army ants possess a discipline, a sense of mission (to judge by appearance, at least), and, collectively, a cohesion, that few human troops attain. At night they bivouac; hundreds of thousands of ants link their bodies together, forming a solid cylinder several feet in diameter, with the queen and her brood secure in the center. At daybreak this body dissolves into temporary chaos, out of which emerges a marching column. The smaller, more agile worker ants take the lead, while bulkier ants, the soldiers, follow along on the shoulder of the trail. Leaves rustle underfoot, and grasshoppers make popping sounds as they dart into tree trunks in self-defense—often to no avail; many of those that succeed in escaping the ants are eaten by ant thrushes, scavenger birds that follow the offensive assiduously from a series of nearby perches.

If you let your eyes fall out of focus, says Wilson, the ants look like a single body, ten or fifteen yards long and more than a yard wide. It grows like a tree, extending and then dividing into two, three, five branches, which subdivide and merge until the ants form a united front, an arc perhaps twenty yards across, that moves methodically forward. This band seeks and destroys tarantulas, scorpions, beetles, roaches, grasshoppers, alien ants—even insufficiently agile snakes and lizards. Death comes through stinging or asphyxiation. Dismemberment and transport proceed without deliberation; half a dozen workers will carry a scorpion's tail back fifteen yards to the "booty cache," from which, by day's end, it will be taken to the bivouac—traveling, all told, the length of a football field.

Less terrifying but no less impressive than the army ants is *Atta sexdens*, a species of agricultural ant. An *Atta* colony employs an intricate division of labor to simultaneously plant, tend, and harvest its staple crop, fungus. The largest workers forage and return to the nest with leaves, which smaller ants cut into manageable sizes. Still smaller ants chew these fragments into little pellets and then turn them over to another caste, whose members embed them in the earth. Upon these beds members of yet another caste implant tiny tufts of fungus. Tending the crop is left to the smallest ants of all; they lick it clean, weed out foreign fungi, and extract strands periodically to feed their sisters.

Although an Atta colony doesn't resemble an organism quite so strikingly as does a colony of army ants, it does possess some of the properties of organisms: division of labor, the selfless devotion of the parts to the whole, and their utter dependence on it. In fact, the same could be said of all ant colonies. William Morton Wheeler, Harvard's great turn-of-the-century entomologist, whose heir E. O. Wilson is, contended that the ant colony is a kind of organism, a "superorganism."

This blurring of the line between society and organism is a delicate matter, and it lies behind some of the criticism to which E. O. Wilson has been exposed. There is something unsettling—to Western sensibilities, especially—about a society that works with such mechanical precision and blind individual sacrifice. Fascist Italy springs immediately to mind, as do Communist China and even modern Japan, which, for all its cheap and trusty cars and computers, is a society whose efficiency some Americans find eerie. Further, it doesn't take much imagination to see a parallel between *Atta*, with its caste system, and a human society in which people are assigned at birth to a permanent socioeconomic status *(as in classical India)*.

Wilson, to be sure, has never held up ant societies as worthy of human emulation, but he has written that they are more nearly "perfect," in a strictly biological sense, than are mammalian societies, and at the left end of the political spectrum—where criticism of Wilson has more than once originated—such language has a reactionary sound to it. It may, then, seem ironic that Wheeler, who took the superorganism idea very literally and seemed to derive a certain aesthetic pleasure from it, was a socialist. But in a way this makes sense. The nagging problem faced by socialist and communist societies has been that humans are selfish; if they are not recompensed in proportion to their work, they tend not to work very hard. An ant society has no such problem. Its altruism is deeply programmed.

The ant colony's unity, in addition to being politically suggestive, is baffling, at least on the face of it, because it is realized in such mindless

E. O. Wilson at a field station of the Organization for Tropical Studies at La Selva, Costa Rica, examining a nest of leafcutting ants (*Atta cephalotes*) in the rain forest.

fashion. Ants are capable of only a small array of stereotyped behaviors and are unable to tell one nestmate from another. How does such individual stupidity add up to societal intelligence?

In Wheeler's day, this was a profound mystery surrounding all socially complex insects—wasps, termites, and bees, as well as ants—and it provoked some fairly weird speculation about immaterial forces, such as "the spirit of the hive." Even Wheeler, in trying to explain the insect societies' coherence, was forced to consider "psychological agencies like consciousness and will." In the case of ants, the correct and more down-to-earth explanation would not be confidently advanced until a half-century after Wheeler wrote. E. O. Wilson, by then a young professor at Harvard, would find the critical piece of evidence while playing with one of his childhood friends from Alabama, *Solenopsis invicta*.

Unraveling the Mystery of Pheromones

In 1953 Wilson attended a lecture by Konrad Lorenz, the Austrian zoologist who founded ethology, commonly defined as the study of animal behavior from an evolutionary perspective. During his lecture, Lorenz talked about "releasers"—signals, usually visual or auditory, that trigger stereotyped sequences of behavior in birds and other animals. Wilson immediately saw the analogous role that chemical signals might play in the social insects.

Several years later, after being diverted by a period of world travel, he tested his hunch. He took a specimen of *Solenopsis invicta* and sacrified it to science. Seeking the source of the odor that emanates from the fire ant's trails, Wilson removed three organs that seemed likely candidates and washed them. (No easy task; each was about the size of a short and inordinately thin piece of thread.) He then crushed each organ into a pulp. Only the Dufour's gland, at the base of the stinger, proved provocative. When its remnants were smeared on a glass pathway to a nest of fire ants, dozens poured out, headed for the spot, and, upon reaching it, milled around as if looking for something to do.

Wilson had found the source of the chemical trail markers that keep fire ants in line as they forage. And he had discovered that Lorenz's releasers are indeed analogous. This chemical—or pheromone, as chemical signals later came to be called—triggers a fairly complex behavioral sequence; it not only keeps ants on the trail but also persuades them to leave the nest and begin the trek in the first place. It is, in Wilson's words, "not just the guidepost, but the entire message." He has written of his reaction to the experiment: "That night I couldn't sleep. I envisioned accounting for the entire social repertory of the ants with a small number of chemical releasers."

He made some progress toward that goal. He isolated the alarm pheromone of the fire ants and harvester ants, which they emit in the face of a threat, sending their compatriots into a combative frenzy. And he found a "necrophoric substance" that accumulates in an ant's body after death, inducing other ants to take the corpse to the graveyard lest it gum up the societal works. Wilson tainted live ants with the substance and watched as they were carted off prematurely, their resistance notwithstanding—an elegant demonstration that ant societies owe a greater debt to the power of pheromones than to the intelligence of ants.

Not all the glory was to be Wilson's. Even as he was discovering the source and effect of trail markers, Martin Lindauer, a German entomologist, was isolating an alarm pheromone in a species of leaf-cutter ant. And Lindauer went further, identifying the compostion of the substance and thus becoming the first scientist to chemically characterize a pheromone.

As for Wilson's dream: it now appears that a fairly small number of signals does indeed account for the complex cohesiveness of ant societies. It is as if each species had a vocabulary of ten or fifteen messages, and everyone took instructions without question. One message means "Follow me." Another: "On guard! A threat to the public good is present." (This was the message in the scent that Wilson had evoked by picking apart the tree stump in Rock Creek Park; the alarm pheromone of *Acanthomyops* is essence of citronella.) Another: "Help me clean my body; it has so many hard-to-reach places." Another: "I'm dead. Get me out of here; I'll only get in the way." There is even evidence that these messages can be combined. When a fire ant is forcibly detained, it emits both alarm and trail pheromone, arousing other ants and pointing them in its direction. In effect: "I'm in trouble and I'm over here."

Kamikaze Bees and Prairie Dog Scouts

The decoding of the messages that bind ant colonies did not send shock waves through the scientific community. Norbert Weiner, the founder of cybernetics, had already established that coordinated behavior of all kinds involves the transmission of information. Besides, cursory inspection of an ant colony suggests that its trails have been invisibly marked; an ant that ventures off the path behaves remarkably like a hound dog that has lost its quarry's scent. So even before Wilson had crushed the Dufour's gland and deciphered its contents, the prevailing suspicion was that ants use some such trail markers and perhaps other chemical signals as well. Pheromones were in the air.

The confirmation of this suspicion, while explaining the unity of ant societies in one sense, left its mystery intact in another. True, the mechanism of orchestration—the medium of communication—had been found. But why were such mechanisms warranted in the first place? Why had evolution built unified, cooperative societies? To put the issue in formal parlance: the proximate cause of insect integration was clear, but the ultimate cause wasn't.

This mystery is deeper than it sounds. Asking why evolution integrated ants so thoroughly is not like asking why it gave them legs or mandibles. Legs and mandibles make immediate evolutionary sense: obviously, they help an ant survive. Social cooperation is not so unequivocally advantageous. Consider the starkest form of cooperation: total self-sacrifice—pure, final, and necessarily unreciprocated altruism. Bees disembowel themselves by stinging intruders. Ants of the species *Camponotus saundersi* defend the homeland by detonating themselves; they contract the muscles of the abdomen so intensely that it ruptures and releases a sticky substance, which turns the ground into flypaper, stymying aggressors. If natural selection really is survival of the fittest, why would it preserve behaviors that are so emphatically not conducive to individual survival?

The issue is not confined to insects. Prairie dogs endanger themselves by conspicuously barking to warn fellow dogs of an approaching coyote or hawk, and the meerkats of Africa also sound alarms upon sighting predators. Many other mammals share food and groom one another, and human beings have been known to jump on hand grenades to save brothers in arms. William Morton Wheeler saw the breadth of selflessness clearly back in 1910. Biologists, he wrote, must not be preoccupied with the "struggle for existence" but rather must explore "the ability of the organism to temporize and compromise with other organisms, to inhibit certain activities of the aequipotential unit in the interests of the unit itself and of other organisms; in a word, to secure survival through a kind of egoistic altruism."

When Wheeler wrote, and for decades thereafter, some biologists considered the explanation for altruism to be not all that tricky. In fact, even today there are people who will dismiss the problem with a wave of hand, casually remarking that altruism evolved for the "good of society." Such remarks should be greeted coolly. One common sign of fuzzy understanding of the theory of natural selection is loose talk about the good of groups. Careful thought shows how difficult it is for an altruistic trait to survive simply on the strength of its benefits to the society, or species, at large.

The way evolution works, so far as we know, is that a new gene or new combination of genes arises—through random mutation or sexual recombination—and somehow improves the chances that an organism will survive and reproduce. That is, the gene brings an "adaptive" trait. But the story is not so simple for genes that benefit organisms other than their own. Imagine a prairie dog colony in which selfishness is rampant, and imagine that a gene for altruism arises. Specifically, suppose the gene inclines the prairie dog to do what prairie dogs in fact do—stand up on its hind legs and sound a warning call upon sighting an invader. How long would such a gene last? Roughly as long as it took its host organism to encounter a coyote. The altruist would dutifully stand up, emit its alarm signal, and, having attracted the invader's attention, get slain and fade into the annals of prairie dog history.

This is not to say that a society full of such individuals wouldn't fare well. On the contrary, it might thrive as no society ever had before. But genetic mutations don't generally appear all across the board; they show up in one of two animals and then have to work their way to wider acceptance. The question, then, is whether a single "warning-call gene," or a handful of them, could pervade the colony in the first place. And the answer appears to be no. There is no obvious reason why altruism should ever get off the ground.

The Theory of Kin Selection

Fortunately for the theory of natural selection, it turns out that there is a way altruism can get off the ground without relying on group selection. It is called "kin selection." The idea behind kin selection is that the gene, not the individual, is the unit of natural selection, and the interests of the gene and the interests of the individual don't always coincide.

If you are a prairie dog (or are just yourself, for that matter) and you have a recently invented gene—synthesized, say, by your great-grandparents—so do roughly half of your siblings and one eighth of your cousins. Now, suppose this gene is a warning-call gene, and suppose some siblings and cousins are in your vicinity when a predator appears. You get up on your hind legs and tip everyone off to the impending danger, in the process tipping the predator off to your location and getting eaten. This may seem like a very valiant thing for you, and your warning-call gene, to do. But, in fact, the "sacrifice" made by your warning-call gene is no such thing.

Sure, the gene perishes along with its "altruistic" possessor (you), but meanwhile, in the bodies of a dozen siblings and cousins, two or three or four or five carbon copies of the gene are carried off safely to be transmitted to future generations. Such a gene will do much better on the evolutionary marketplace than a "coward" gene, which would save itself only to see its several replicas plucked from the gene pool. The theory of kin selection is very much in the spirit of Samuel Butler's observation that "a hen is only an egg's way of making another egg." Only it adds this point: "So is that hen's sister."

It is important to be clear on what the theory of kin selection does and doesn't say. It doesn't say that genes can sense copies of themselves in another organism and direct their own organism to behave hospitably toward it. Genes are not clairvoyant, and they are not little puppet masters that govern behavior on a day-to-day basis. Their main influence on behavior comes through their construction of the brain, which thereafter is in charge. The theory of kin selection says simply that natural selection can be expected to permit the proliferation of genes that build brains that lead to behaviors that are likely to help kin.

Anyone who understands this clearly should thrill at the subtlety and power of the theory of natural selection, of which the theory of kin selection is a corollary.

The theory of kin selection changed the mathematical convention of evolutionists. Now the models of population biology are based on the assumption that evolution maximizes not "individual fitness" but "inclusive fitness." This term is broad enough to encompass the gene's total contribution to the survival and reproduction of the information encoded in it, regardless of whether the immediate beneficiary is its particular vehicle or a different vehicle containing the same information. The new math can be used to show how food sharing, grooming, and various valorous behaviors could have a genetic basis, and to define the theoretical limits of such altruism. This logic has been summed up in an anecdote about an evolutionary biologist *(J. B. S. Haldane)* who was asked whether, after all he had learned about the ruthless genetic calculus programmed into people, he would still, as he had once vowed, give his life for his brother. "No," he is said to have replied. "Two brothers or eight cousins."

The theory of kin selection, though implicit in the writings of Darwin, was not cast in mathematical form (nor widely appreciated even in verbal form) until 1964. In that year, William D. Hamilton developed it rigorously and used it to solve—or, more accurately, to suggest a very plausible solution to—the puzzle of the social insects' superorganic unity. As Wilson once summed up this achievement before a class of undergraduates at Harvard, "Now it was a remarkable achievement in the early sixties by a then-young British biologist named Bill Hamilton, who looked at social insects in a wholly new way—and I might add, just for your delectation, a footnote. At that time he and I were probably the leading students, younger students, of social insects. And he looked at it in a way that would have never crossed my mind."

Hamilton, like other entomologists, was perplexed by insects of the order Hymenoptera—wasps, bees, and ants. Why is it, he wondered, that highly integrated social behavior is found in virtually no insect groups outside of Hymenoptera? (Termites are the lone exception.) What do wasps, bees, and ants have in common that would account for their cohesion?

One thing they have in common is an odd approach to reproduction *(called haplodiploidy)*. Some of their eggs yield life without fertilization. In fact, what determines the sex of an unborn ant is whether its egg is fertilized. If it is, then it becomes a female; if not—if it receives no genetic input from the male—then it becomes, ironically enough, a male. This means that when one of these males matures and produces sperm, the genetic information is the same in all of his sperm cells; having come from an unfertilized egg, he has only one set of chromosomes *(haploid)* on which to draw. The queen, in contrast, having emerged

A queen of the leafcutter ant *Atta cephalotes* surrounded by her workers, all of whom are her daughters. The different sizes represent different castes, which perform special functions in the colony. This is a species studied especially thoroughly by E. O. Wilson.

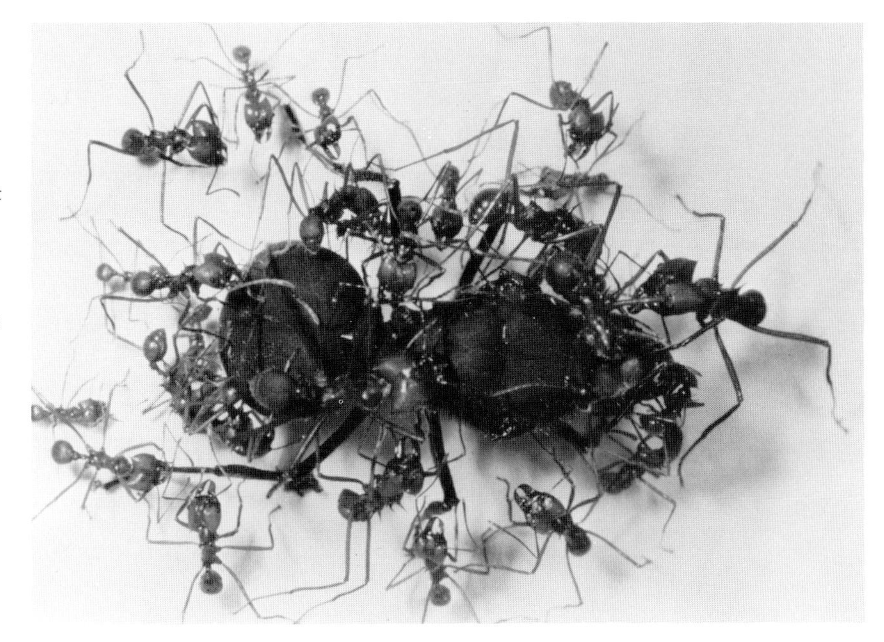

from a fertilized egg, will have (like humans) two sets of chromosomes *(diploid)*—twice as much genetic information as was actually needed to construct her. Since each egg she produces will draw randomly on this store, her eggs will differ from one another.

So ant eggs are just like most eggs in the animal kindgom; any two produced by the same mother have about half of their genes in common. But ant sperm are weird; any two sperm cells produced by the same father are identical. The upshot is this: the various females born of the fusion of these sperm cells and these eggs will be very closely related—much more so than ordinary "sisters."

Whereas two human sisters have about one half of their genes in common, two ant sisters share, on average, three fourths of their genes. Actually, these fractions are misleading. People in fact share much more than half of their genes with *any* given person—and, for that matter, with any given chimpanzee. But fairly novel genes, genes that are not yet established in the population, do indeed have a 50 percent chance of residing in the sisters of their human carriers and a 75 percent chance of residing in the sisters of their ant carriers. And novel genes, being on the cutting edge of evolution, are the ones we're interested in here.

In his 1964 paper, Hamilton argued that the large genetic overlap among ants revises the mathematics of altruism. If it makes genetic sense for a prairie dog to die because otherwise four fertile sisters will each face a 51 percent chance of death, he reasoned, the same sacrifice makes sense for an ant with only three similarly threatened fertile sisters. If it makes sense for a monkey to share a little food with her sister monkeys, it makes sense for an ant to share a lot of food. In general, the logic appears to be slanted toward cooperation and self-sacrifice in the order Hymenoptera. The line between self-interest and the interest of a sibling, always a little blurry, is blurred further in ant, wasp, and bee societies.

Wilson has told his students,

> Hamilton said, "Is it possible that this bias, this kin selection, is the driving force behind repeated origination of this type of higher colonial organization in the bees, wasps, and ants?" And he sold me immediately on this idea. And I was one of the early—I was *the* early defender of this view. And I must say that I've had to concede that Hamilton—even though I think I knew more about social insects—Hamilton beat me to it to produce the main idea, the most original, important idea on social insects of this century.
>
> I had to react the way young Huxley, Thomas Henry Huxley, reacted when he read *The Origin of Species*. Here was Darwin saying, Look, we can explain all these marvelous things by natural selection. . . . And Huxley's comment was the one that I made: How stupid of me not to have thought of that. Why didn't I sit down and think for a few minutes instead of running out in the field and, you know, doing all these things? Well, anyway, as a consequence, kin selection, and this basic approach, has become central in the development of the field of sociobiology.

A Reductionist—And Proud of It!

More than most scientists, E. O. Wilson thinks about science. He thinks about the rules of conduct, the drive to discover, the social organization of the enterprise and its conceptual organization. And there is one image, implicit in his thinking, that almost single-handedly accounts for his work over the past dozen years, with all its ups and downs—the Pulitzer Prize, the National Medal of Science, the alliances formed and the friendships broken, the praise that he relishes and the criticism that has hurt him. This is the image of a solid structure rising certainly into the air—a tower, or a skyscraper, or, perhaps, a pyramid. At the top of the pyramid are the social sciences. Below them are the biological sciences. Below them is chemistry, and below it is physics.

It is not a new metaphor, by any means; it has occurred to thousands of thinkers over the years. But Wilson takes it more seriously than most. As science progresses, he believes, each level of inquiry will be seen to rest on the level beneath it in a fairly literal sense: its laws will follow from the laws below, almost as surely as the Pythagorean theorem follows from the basic assumptions of plane geometry.

This belief is known as reductionism. At the lower levels of organization, reductionism will not get you into hot water. Many of the laws of chemistry have been reduced in a rigorous way to the laws of physics, so reductionism is not, among chemists, an arguable philosophy. (That is not to say they are enthusiastic about it. Because of this reduction, some of the work they once did with test tubes can now be done by mathematicians with a knowledge of physics.)

It is only in the higher regions of the pyramid of knowledge that reductionism becomes disputable. Does biology literally rest on chemistry? Could the behavior of, say, a kidney be predicted with much precision from a knowledge of the molecules involved? How about a brain? Could laws describing a dog's or a chimpanzee's or a person's behavior be deduced from the laws of organic chemistry?

These questions, in addition to being difficult, are loaded with philosophical consequence. At these higher levels of organization, reductionism is allied with determinism, which holds that free will is a myth, that the rest of human history will unfold inevitably, however powerful the illusion that we are directing it with our various "choices." Our inability to predict this predetermined future—or even to predict one person's behavior on a day-to-day basis—reflects, according to determinists, only incomplete data and our ignorance of the principles involved.

After all, humans are very complicated; their behavior is guided by millions of microscopic signals per split second. In practice, the action is simply impossible to keep track of. Nonetheless, if each signal is, as we presume, the inevitable result of some other signal or signals, which in turn were grounded with equal firmness in their antecedents, then we could, in principle, project the path of behavioral causality forward and show that even the most ethereal and inspiring of human accomplishments is the product of mechanical necessity. Thus say the determinists, and thus say the reductionists. And thus are they not the life of many parties.

Further besmirching the reputation of reductionists at the pinnacle of the pyramid of knowledge is the fact that social scientists, like chemists, are not eager to surrender turf. And the less eager they are, the more inclined they are to see imperialistic designs in the ideas that drift up from below. Thus, within the social sciences, the word reductionism has taken on distinctly pejorative, and offensive, connotations: a reductionist is someone who offers simplistic biological explanations of human behavior, often with hostile intent. E. O. Wilson, though once insensitive to the mores of the social sciences, has become intimately and painfully acquainted with these connotations since 1975, when his book *Sociobiology* was published.

Whether Wilson really is a reductionist in this sense of the word is still a subject of debate. But he is without doubt a reductionist in the pure sense of the word—and proud of it. Since 1981, in fact, he has been pushing a theory that is just about as reductionist in spirit as a theory can be. It is an attempt to encompass in one fell swoop most of the pyramid of science, an attempt to formulate laws linking molecular biology to psychology, anthropology, and sociology. Needless to say, this theory—the "gene-culture theory," set forth in a book called *Genes, Mind, and Culture*, which Wilson coauthored with a young physicist named Charles Lumsden—has not found favor with truckloads of social scientists. Many biologists are also skeptical of it. Further, and more significantly, a number of sociobiologists are unhappy with the theory.

The problem, as Wilson diagnosed it, was that people didn't understand. It wasn't so much that they didn't understand the gene-culture theory—though that was part of the problem—as that they didn't understand science. Science, he says, is a continual oscillation between expansion and compression. The cycle begins when a scientist—a "systems builder," more specifically—comes along and claims he can compress a large body of information into a small theoretical package. The gene-culture theory, for example, is an attempt to account for masses of data about human cultural history with a fairly small set of principles. It may not immediately be clear how a theory is going to accommodate all the information in its domain; being new, it is still rudimentary. But as it is tested and refined, it will begin to "unfold into a full display of rich detail."

In the meantime, though, you can't expect the theory's creator to sit around contemplating its shortcomings. That is left to the "bookkeepers," the scientists who "fill in the blanks." It is their job to collect lots of data and see whether they really do fall into the pattern predicted by the theory.

If you're a bookkeeper, says Wilson, "you can afford to play by the rules and express lack of confidence in your results and express the severe doubts that the experimental data may engender." Wilson used to do a lot of bookkeeping, but these days he sticks mainly to synthesis and systems building. And when you're a systems builder he says, "you have to have something like faith. You have to believe as you go on that in fact there is some major organizing principle that remains to be discovered. You have to believe that indeed it exists and that no matter how imperfect a formulation may be from one year to the next, or what setbacks might occur, how many seemingly intractable methodological problems originate, that this will prove to be correct."

"It was in college that I came up against it in its full grandeur," says Wilson of the theory of natural selection. "I was completely taken by that as an organizing principle." He was struck not just by the power of the principle but by the legitimacy it gave his career plans. The idea that everything about every animal could be accounted for with "a real, scientific, unifying explanation was totally transforming for me."

I once asked Wilson if the word epiphany could be accurately applied to such intellectual experiences. "I think there probably is a similar emotional response," he said. "You know, in the typical epiphany, or conversion, the individual says something along the lines of 'I discovered

God. Jesus came into my life.' But the outcome of all this is that the individual sees the unity in the universe. Instead of just this fragmented world in which he's doing selfish acts, he sees a purpose for the universe of which he is a small part. And in a tribalistic manner he now submerges himself into the grand plan, a great plan. And that brings a certain very profound peace. Now in a somewhat related way—but with real differences—I think that discovering something of a unifying idea gives you a sense that you do have the key, not to the universe but to the big chunk of it that matters the most to you. In my case, what mattered the most to me was biological diversity."

*Edward O. Wilson, a Harvard professor, has written or cowritten half a dozen books on various aspects of sociobiology, including one popular book that garnered a Pulitzer Prize (*On Human Nature*, 1978). As a nontechnical summary, however, the article that appeared in the 12 October 1975* New York Times Magazine *is one of his best. The essay "Human Decency is Animal" by Edward O. Wilson (copyright 1975 by the New York Times Company) is reprinted here by permission of the author and the Times.*

Why Sociobiology?

Edward O. Wilson

During the American wars of this century, a large percentage of Congressional Medals of Honor were awarded to men who threw themselves on top of grenades to shield comrades, aided the rescue of others from battle sites at the price of certain death to themselves, or made other, often carefully considered but extraordinary, decisions that led to the same fatal end. Such altruistic suicide is the ultimate act of courage and emphatically deserves the country's highest honor. It is also only the extreme act that lies beyond the innumerable smaller performances of kindness and giving that bind societies together.

One is tempted to leave the matter there, to accept altruism as simply the better side of human nature. Perhaps, to put the best possible construction on the matter, conscious altruism is a transcendental quality that distinguishes human beings from animals. Scientists are nevertheless not accustomed to declaring any phenomenon off limits, and recently there has been a renewed interest in analyzing such forms of social behavior in greater depth and as objectively as possible.

Much of the new effort falls within a discipline called sociobiology, which is defined as the systematic study of the biological basis of social behavior in every kind of organism, including man. It is being pieced together with contributions from biology, psychology and anthropology. There is of course nothing new about analyzing social behavior, and even the word sociobiology has been around for some years. What is new is the way facts and ideas are being extracted from their traditional matrix of psychology and ethology (the natural history of animal behavior) and reassembled in compliance with the principles of genetics and ecology.

Iwo Jima Memorial

In sociobiology, there is a heavy emphasis on the comparison of societies of different kinds of animals and of man, not so much to draw analogies (these have often been dangerously misleading, as when aggression is compared directly in wolves and in human beings) but to devise and to test theories about the underlying hereditary basis of social behavior. With genetic evolution always in mind, sociobiologists search for the ways in which the myriad forms of social organization adapt particular species to the special opportunities and dangers encountered in their environment.

Altruism in Mammals and Birds

A case in point is altruism. I doubt if any higher animal, such as a hawk or a baboon, has ever deserved a Congressional Medal of Honor by the ennobling criteria used in our society. Yet minor altruism does occur frequently, in forms instantly understandable in human terms, and is bestowed not just on offspring but on other members of the species as well. Certain small birds, robins, thrushes and titmice, for example, warn others of the approach of a hawk. They crouch low and emit a distinctive thin, reedy whistle. Although the warning call has acoustic properties that make it difficult to locate in space, to whistle at all seems at the very least unselfish; the caller would be wiser not to betray its presence but rather to remain silent and let someone else fall victim.

When a dolphin is harpooned or otherwise seriously injured, the typical response of the remainder of the school is to desert the area immediately. But, sometimes, they crowd around the stricken animal and lift it to the surface, where it is able to continue breathing air. Packs of African wild dogs, the most social of all carnivorous mammals, are organized in part by a remarkable division of labor. During the denning season, some of the adults, usually led by a dominant male, are forced to leave the pups behind in order to hunt for antelopes and other prey. At least one adult, normally the mother of the litter, stays behind as a guard. When the hunters return, they regurgitate pieces of meat to all that stayed home. Even sick and crippled adults are benefited, and as a result they are able to survive longer than would be the case in less generous societies.

Other than man, chimpanzees may be the most altruistic of all mammals. Ordinarily, chimps are vegetarians, and during their relaxed foraging excursions they feed singly in the uncoordinated manner of other monkeys and apes. But, occasionally, the males hunt monkeys and young baboons for food. During these episodes, the entire mood of the troop shifts toward what can only be characterized as a manlike state. The males stalk and chase their victims in concert; they also gang up to repulse any of the victim's adult relatives which oppose them. When the hunters have dismembered the prey and are feasting, other chimps approach to beg for morsels. They touch the meat and the faces of the males, whimpering and hooing gently, and holding out their hands—palms up—in supplication. The meat eaters sometimes pull away in refusal or walk off. But, often, they permit the other animal to chew directly on the meat or to pull off small pieces with its hands. On several occasions, chimpanzees have actually been observed to tear off pieces and drop them into the outstretched hands of others—an act of generosity unknown in other monkeys and apes.

Adoption is also practiced by chimpanzees. Jane Goodall has observed three cases at the Gombe Stream National Park in Tanzania. All involved orphaned infants taken over by adult brothers and sisters. It is of considerable interest, for more theoretical reasons to be discussed shortly, that the altruistic behavior was displayed by the closest possible relatives rather than by experienced females with children of their own, females who might have supplied the orphans with milk and more adequate social protection.

Insect Altruism

In spite of a fair abundance of such examples among vertebrate creatures, it is only in the lower animals and in the social insects particularly, that we encounter altruistic suicide comparable to man's. A large percentage of the members of colonies of ants, bees, and wasps are ready to defend their nests with insane charges against intruders. This is the reason that people move with circumspection around honeybee hives and yellowjacket burrows, but can afford to relax near the nests of solitary species such as sweat bees and mud daubers.

The social stingless bees of the tropics swarm over the heads of human beings who venture too close, locking their jaws so tightly onto tufts of hair that their bodies pull loose from their heads when they are combed out. Some of the species pour a burning glandular secretion onto the skin during these sacrificial attacks. In Brazil, they are called "cagafogos" (fire defecators). The great entomologist William Morton Wheeler described an encounter with the "terrible bees," during which they removed patches of skin from his face, as the worst experience of his life.

Honeybee workers have stings lined with reversed barbs like those on fishhooks. When a bee attacks an intruder at the hive, the sting catches in the skin; as the bee moves away, the sting remains embedded, pulling out the entire venom gland and much of the viscera with it. The bee soon dies, but its attack has been more effective than if it withdrew the sting intact. The reason is that the venom gland continues to leak poison into the wound, while a bananalike odor emanating from the base of the sting incites other members of the hive into launching Kamikaze attacks of their own at the same spot. From the point of view of the colony as a whole, the suicide of an individual accomplishes more than it loses. The total worker force consists of 20,000 to 80,000 members, all sisters born from the eggs laid by the mother queen. Each bee has a natural life span of only about 50 days, at the end of which it dies of old age. So to give a life is only a little thing, with no genes being spilled in the process.

Explaining Altruism through Kin Selection

Sharing a capacity for extreme sacrifice does not mean that the human mind and the "mind" of an insect (if such exists) work alike. But it does mean that the impulse need not be ruled divine or otherwise transcendental, and we are justified in seeking a more conventional biological explanation. One immediately encounters a basic problem connected with such an explanation. Fallen heroes don't have any more children.

If self-sacrifice results in fewer descendants, the genes, or basic units of heredity, that allow heroes to be created can be expected to disappear gradually from the population. This is the result of the narrow mode of Darwinian natural selection: Because people who are governed by selfish genes prevail over those with altruistic genes, there should be a tendency over many generations for selfish genes to increase in number and for the human population as a whole to become less and less capable of responding in an altruistic manner.

How can altruism persist? In the case of the social insects, there is no doubt at all. Natural selection has been broadened to include a process called kin selection. The self-sacrificing termite soldier protects the rest of the colony, including the queen and king which are the soldier's parents. As a result, the soldier's more fertile brothers and sisters flourish, and it is they which multiply the altruistic genes that are shared with the soldier by close kinship. One's own genes are multiplied by the greater production of nephews and nieces.

It is natural, then, to ask whether the capacity for altruism has also evolved in human beings through kin selection. In other words, do the emotions we feel, which on occasion in exceptional individuals climax in total self-sacrifice, stem ultimately from hereditary units that were implanted by the favoring of relatives during a period of hundreds or thousands of generations? This explanation gains some strength from the circumstance that during most of mankind's history the social unit was the immediate family and a tight network of other close relatives. Such exceptional cohesion, combined with a detailed awareness of kinship made possible by high intelligence, might explain why kin selection has been more forceful in human beings than in monkeys and other mammals.

On Human Nature

To anticipate a common objection raised by many social scientists and others, let me grant at once that the intensity and form of altruistic acts are to a large extent culturally determined. Human social evolution is obviously more cultural than genetic. The point is that the underlying emotion, powerfully manifested in virtually all human societies, is what is considered to evolve through genes. This sociobiological hypothesis does not therefore account for differences among societies, but it could explain why human beings differ from other mammals and why, in one narrow aspect, they more closely resemble social insects.

In cases where sociobiological explanations can be tested and proved true, they will, at the very least, provide perspective and a new sense of philosophical ease about human nature. I believe that they will also have an ultimately moderating influence on social tensions. Consider the case of homosexuality. Homophiles are typically rejected in our society because of a narrow and unfair biological premise made about them: Their sexual preference does not produce children; therefore, they cannot be natural. To the extent that this view can be rationalized, it is just Darwinism in the old narrow sense: Homosexuality does not directly replicate genes. But homosexuals can replicate genes by kin selection provided they are sufficiently altruistic toward kin.

It is not inconceivable that in the early, hunter-gatherer period of human evolution, and perhaps even later, homosexuals regularly served as a partly sterile caste, enhancing the lives and reproductive success of

Mosaic II (M. C. Escher, 1957)

their relatives by a more dedicated form of support than would have been possible if they produced children of their own. If such combinations of interrelated heterosexuals and homosexuals regularly left more descendants than similar groups of pure heterosexuals, the capacity for homosexual development would remain prominent in the population as a whole. And it has remained prominent in the great majority of human societies, to the consternation of anthropologists, biologists and others.

Supporting evidence for this new kin-selection hypothesis does not exist. In fact, it has not even been examined critically. But the fact that it is internally consistent and can be squared with the results of kin selection in other kinds of organisms should give us pause before labeling homosexuality an illness. I might add that if the hypothesis is correct, we can expect homosexuality to decline over many generations. The reason is that the extreme dispersal of family groups in modern industrial societies leaves fewer opportunities for preferred treatment of relatives. The labor of homosexuals is spread more evenly over the population at large, and the narrower form of Darwinian natural selection turns against the duplication of genes favoring this kind of altruism.

On Aggression

A peacemaking role of modern sociobiology also seems likely in the interpretation of aggression, the behavior at the opposite pole from altruism. To cite aggression as a form of social behavior is, in a way, contradictory considered by itself. It is more accurately identified as anti-social behavior. But, when viewed in a social context, it seems to be one of the most important and widespread organizing techniques. Animals use it to stake out their own territories and to establish their rank in the pecking orders. And because members of one group often cooperate for the purpose of directing aggression at competitor groups, altruism and hostility have come to be opposite sides of the same coin.

Konrad Lorenz, in his celebrated book *On Aggression*, argued that human beings share a general instinct for aggressive behavior with animals, and that this instinct must somehow be relieved, if only through competitive sport. Erich Fromm, in *The Anatomy of Human Destructiveness*, took the still dimmer view that man's behavior is subject to a unique death instinct that often leads to pathological aggression beyond that encountered in animals. Both of these interpretations are essentially wrong. A close look at aggressive behavior in a variety of animal societies, many of which have been carefully studied only since the time Lorenz drew his conclusions, shows that aggression occurs in a myriad of forms and is subject to rapid evolution.

We commonly find one species of bird or mammal to be highly territorial, employing elaborate, aggressive displays and attacks, while a second, otherwise similar species shows little or no territorial behavior.

In short, the case for a pervasive aggressive instinct does not exist. The reason for the lack of a general drive seems quite clear. Most kinds of aggressive behavior are perceived by biologists as particular responses to crowding in the environment. Animals use aggression to gain control over necessities—usually food or shelter—which are in short supply or likely to become short at some time during the life cycle. Many species seldom, if ever, run short of these necessities; rather, their numbers are controlled by predators, parasites, or emigration. Such animals are characteristically pacific in their behavior toward one another.

Mankind, let me add at once, happens to be one of the more aggressive species. But we are far from being the most aggressive. Recent studies of hyenas, lions and langur monkeys, to take three familiar species, have disclosed that under natural conditions those animals engage in lethal fighting, infanticide and even cannibalism at a rate far above that found in human beings. When a count is made of the number of murders committed per thousand individuals per year, human beings are well down the list of aggressive creatures, and I am fairly confident that this would still be the case even if our episodic wars were to be averaged in.

Alongside ants, which conduct assassinations, skirmishes and pitched battles as routine business, men are all but tranquil pacifists. Ant wars, incidentally, are especially easy to observe during the spring and summer in most towns and cities in the Eastern United States. Look for masses of small, blackish brown ants struggling together on sidewalks or lawns. The combatants are members of rival colonies of the common pavement ant, *Tetramorium caespitum*. Thousands of individuals may be involved, and the battlefield typically occupies several square feet of the grassroots jungle.

Despite the fact that many kinds of animals are capable of a rich, graduated repertory of aggressive actions, and despite the fact that aggression is important in the organization of their societies, it is possible for individuals to go through a normal life, rearing offspring, with nothing more than occasional bouts of playfighting and exchanges of lesser hostile displays. The key is the environment: Frequent intense display and escalated fighting are adaptive responses to certain kinds of social stress which a particular animal may or may not be fortunate enough to avoid during its lifetime. By the same token, we should not be surprised to find a few human cultures, such as the Hopi or the newly discovered Tasaday of Mindanao *(since alleged to be an imposter group of modern individuals attempting to fool anthropologists, perhaps to achieve protection of wildlands from development)*, in which aggressive interactions are minimal. In a word, the evidence from comparative studies of animal behavior cannot be used to justify extreme forms of aggression, bloody drama or violent competitive sports practiced by man.

Nature or Nurture?

This brings us to the topic which, in my experience, causes the most difficulty in discussions of human sociobiology: the relative importance of genetic versus environmental factors in the shaping of behavioral traits. I am aware that the very notion of genes controlling behavior in human beings is scandalous to some scholars. They are quick to project the following political scenario: Genetic determinism will lead to support for the status quo and continued social injustice. Seldom is the equally plausible scenario considered: Environmentalism *(not protection of the environment, but the belief that one's social environment is the most powerful determinant of human behavior)* will lead to support for authoritarian mind control and worse injustice. Both sequences are highly unlikely, unless politicians or ideologically committed scientists are allowed to dictate the uses of science. Then anything goes.

That aside, concern over the implications of sociobiology usually proves to be due to a simple misunderstanding about the nature of heredity. Let me try to set the matter straight as briefly but fairly as possible. What the genes prescribe is not necessarily a particular behavior but the capacity to develop certain behaviors and, more than that, the tendency to develop them in various specified environments. It is the evolution of this pattern which sociobiology attempts to analyze.

We can be more specific about human patterns. It is possible to make a reasonable inference about the most primitive and general human social traits by combining two procedures. First, note is made of the most widespread qualities of hunter-gatherer societies. Although the behavior of the people is complex and intelligent, the way of life to which their cultures are adapted is primitive. The human species evolved with such an elementary economy for hundreds of thousands of years; thus, its innate pattern of social responses can be expected to have been principally shaped by this way of life. The second procedure is to compare the most widespread hunter-gatherer qualities with similar behavior displayed by the species of langurs, colobus, macaques, baboons, chimpanzees, gibbons and other Old World monkeys and apes that, together, comprise man's closest living relatives.

Where the same pattern of traits occurs in man—and in most or all of the primates—we conclude that it has been subject to relatively little evolution. Its possession by hunter-gatherers indicates (but does not prove) that the pattern was also possessed by man's immediate ancestors; the pattern also belongs to the class of behaviors least prone to change even in economically more advanced societies. On the other hand, when the behavior varies a great deal among the primate species, it is less likely to be resistant to change.

Prehistoric rock painting, Australia

Genes and Gender Roles

The list of basic human patterns that emerges from this screening technique is intriguing:

- The number of intimate group members is variable but normally 100 or less.
- Some amount of aggressive and territorial behavior is basic, but its intensity is graduated and its particular forms cannot be predicted from one culture to another with precision.
- Adult males are more aggressive and are dominant over females.
- The societies are to a large extent organized around prolonged maternal care and extended relationships between mothers and children.
- Play, including at least mild forms of contest and mock-aggression, is keenly pursued and probably essential to normal development.

We must then add the qualities that are so distinctively, ineluctably human that they can be safely classified as genetically based: the overwhelming drive of individuals to develop some form of a true, semantic language, the rigid avoidance of incest by taboo and the weaker but still strong tendency for sexually bonded women and men to divide their labor into specialized tasks.

In hunter-gatherer societies, men hunt and women stay at home. This strong bias persists in most agricultural and industrial societies and, on that ground alone, appears to have a genetic origin. No solid evidence exists as to when the division of labor appeared in man's ancestors or how resistant to change it might be during the continuing revolution for women's rights. My own guess is that the genetic bias is intense enough to cause a substantial division of labor even in the most free and most egalitarian of future societies.

As shown by research recently summarized in the book *The Psychology of Sex Differences*, by Eleanor Emmons Maccoby and Carol Nagy Jacklin, boys consistently show more mathematical and less verbal ability than girls on the average, and they are more aggressive from the first hours of social play at age two to manhoood. Thus, even with identical education and equal access to all professions, men are likely to continue to play a disproportionate role in political life, business and science. But that is only a guess and, even if correct, could not be used to argue for anything less than sex-blind admission and free personal choice.

Certainly, there are no a priori grounds for concluding that the males of a predatory species must be a specialized hunting class. In chimpanzees, males are the hunters; which may be suggestive in view of the fact that these apes are by a wide margin our closest living relatives. But, in lions, the females are the providers, typically working in groups with their cubs in tow. The stronger and largely parasitic males hold back

Les Desmoiselles d'Avignon (Pablo Picasso, 1907)

from the chase, but rush in to claim first share of the meat when the kill has been made. Still another pattern is followed by wolves and African wild dogs: Adults of both sexes, which are very aggressive, cooperate in the hunt.

Sociobiology as a Tool for Change

The moment has arrived to stress that there is a dangerous trap in sociobiology, one which can be avoided only by constant vigilance. The trap is the "naturalistic fallacy" of ethics, which uncritically concludes that what is, should be. The "what is" in human nature is to a large extent the heritage of a Pleistocene, hunter-gatherer existence. When any genetic bias is demonstrated, it cannot be used to justify a continuing practice in present and future societies. Since most of us live in a radically new environment of our own making, the pursuit of such a practice would be bad biology; and like all bad biology, it would invite disaster. For example, the tendency under certain conditions to conduct warfare against competing groups might well be in our genes, having been advantageous to our Neolithic ancestors, but it could lead to global suicide now. To rear as many healthy children as possible was long the road to security; yet with the population of the world brimming over, it is now the way to environmental disaster.

Our primitive old genes will therefore have to carry the load of much more cultural change in the future. To an extent not yet known, we trust—we insist—that human nature can adapt to more encompassing forms of altruism and social justice. Genetic biases can be trespassed, passions averted or redirected, and ethics altered; and the human genius for making contracts can continue to be applied to achieve healthier and freer societies. Yet the mind is not infinitely malleable. Human sociobiology should be pursued and its findings weighed as the best means we have of tracing the evolutionary history of the mind. In the difficult journey ahead, during which our ultimate guide must be our deepest and, at present, least understood feelings, surely we cannot afford an ignorance of history.

Evolution and Ethics

Fellow sociobiologist Richard D. Alexander expands on this last point in his book The Biology of Moral Systems *(New York: Aldine de Gruyter, copyright 1987 by Richard D. Alexander, reprinted here by permission of the author).*

Richard D. Alexander

During 1977–78, I participated in a series of conferences at the Hastings Institute in New York, charged with a search for the "foundations" of ethics. These conferences were organized as a result of the flood of new ethical questions arising out of technology: abortion, euthanasia, the right to reject certain kinds of technological prolongation of life, etc.

The Hastings Institute itself was organized for the same general reason, in 1969, and it has become widely known for its examination of the questions involving medical ethics and bioethics. More than anything else, the Hastings conferences set me to thinking and reading about the general background of human ideas about morality. As a result, I began to puzzle over what I saw as an incompleteness or inconclusiveness, not only of the collection of Hastings reports and the discussions in the conferences in which I had participated, but, indeed, of all general arguments on ethics.

Even if quick and decisive solutions to ethical and moral problems are simply not possible, I find myself thinking that in all of the arguments something is missing. Indeed, I believe that something crucial has been missing from all of the great debates of history, among philosophers, politicians, theologians, and thinkers from other and diverse backgrounds, on the issues of morality, ethics, justice, right and wrong. Why have the greatest minds throughout history left such questions seemingly as unresolved as ever? Why should it be that, despite our most intense and sincere efforts, we are not really prepared to deal with the crises that our own activities bring about, even though we cannot but admit that they now have the potential, incredibly, to cause the destruction of life, or at least civilization, on our planet?

Part of the answer is that those who have tried to analyze morality have failed to treat the human traits that underlie moral behavior as outcomes of evolution—as outcomes of the process, dominated by natural selection, that forms the organizing principle of modern biology. This omission is not really the fault of those who have studied ethical and moral systems. The science of biology has been so vague about how to apply natural selection that not until the last two decades have biologists themselves been able to use it to analyze social systems.

Biology provides a broad source of information about humans that has no substitute. It clarifies long-standing paradoxes. It shows that some things have indeed been missing from the debates about morality, and they have been missing because the process of organic evolution that gave rise to all forms of life has been left out of the discussions. Knowledge of the human background in organic evolution can provide a deeper self-understanding by an increasing proportion of the world's population; self-understanding that I believe can contribute to answering the serious questions of social living.

13 *Not in Our Genes*

Harsh criticism of sociobiology is presented by three scientists who collaborated on the 1984 book Not in Our Genes. *Richard C. Lewontin is one of today's leading theorists and experimentalists in the field of evolutionary genetics. Like Edward O. Wilson, he teaches at Harvard University. Steven Rose is head of the neurobiology department at The Open University in Britain and author of several books, including* The Conscious Brain. *Leon J. Kamin is chairman of the psychology department at Northeastern University in Boston. His 1974 book,* The Science and Politics of IQ, *was the first to expose the fraud of Cyril Burt's scientific research and writing on the inheritance of intelligence. Excerpts are drawn from* Not in Our Genes: Biology, Ideology, and Human Nature, *by R. C. Lewontin, Steven Rose, and Leon J. Kamin (copyright 1984 by R. C. Lewontin, Steven Rose, and Leon J. Kamin and reprinted by permission of Pantheon Books, a division of Random House, Inc.).*

Radical Science versus the New Right

Lewontin, Rose, and Kamin

The authors of *Not in Our Genes* are respectively an evolutionary geneticist, a neurobiologist, and a psychologist. Over the past decade and a half we have watched with concern the rising tide of biological determinist writing, with its increasingly grandiose claims to be able to locate the causes of the inequalities of status, wealth, and power between classes, genders, and races in Western society in a reductionist theory of human nature. Each of us has been engaged for much of this time in research, writing, speaking, teaching, and public political activity in opposition to the oppressive forms in which determinist ideology manifests itself. We share a commitment to the prospect of the creation of a more socially just—a socialist—society. And we recognize that a critical science is an integral part of the struggle to create that society, just as we also believe that the social function of much of today's science is to hinder the creation of that society by acting to preserve the interests of the dominant class, gender, and race. This belief—in the possibility of a critical and liberatory science—is why we have each in our separate ways and to varying degrees been involved in the development of what has become known over the 1970s and 1980s, in the United States and Britain, as the radical science movement.

The need was, we felt, for a systematic exploration of the scientific and social roots of biological determinism, an analysis of its present-day social functions, and an exposure of its scientific pretensions. More than that, though, it was also necessary to offer a perspective on what biology and psychology can offer as an alternative, a liberatory, view of the "nature of human nature." Hence, *Not in Our Genes*.

The start of the decade of the 1980s was symbolized, in Britain and the United States, by the coming to power of new conservative governments; and the conservatism of Margaret Thatcher and Ronald Reagan marks in many ways a decisive break in the political consensus of liberal conservatism that has characterized governments in both countries for the previous twenty years or more. It represents the expression of a newly coherent and explicitly conservative ideology often described as the New Right. The response of the liberal consensus to challenges to its institutions has always been the same: an increase in interventive programs of social amelioration; of projects in education, housing, and inner-city renewal. By contrast, the New Right diagnoses the liberal medicine as merely adding to the ills by progressively eroding the "natural" values that had characterized an earlier phase of capitalist industrial society.

But New Right ideology goes further than mere conservatism and makes a decisive break with the concept of an organic society whose members have reciprocal responsibilities. Underlying its "cri de coeur" is a philosophical tradition of individualism, with its emphasis on the priority of the individual over the collective. And the roots of this methodological individualism lie in a view of human nature which it is the main purpose of this book to challenge.

Challenging the Hobbesian Worldview

Philosophically this view of human nature is very old; it goes back to the emergence of bourgeois society in the seventeenth century and to Hobbes's view of human existence as a "bellum omnium contra omnes," a war of all against all, leading to a state of human relations manifesting competitiveness, mutual fear, and the desire for glory. For Hobbes, it followed that the purpose of social organization was merely to regulate these inevitable features of the human condition. And Hobbes's view of the human condition derived from his understanding of human biology; it was biological inevitability that made humans what they were. Such a belief encapsulates the twin philosophical stances with which this book is concerned.

The first is 'reductionism'—the name given to a set of general methods and modes of explanation both of the world of physical objects and of human societies. Broadly, reductionists try to explain the properties of complex wholes—molecules, say, or societies—in terms of the units of which those molecules or societies are composed. They would argue, for example, that the properties of a protein molecule could be uniquely determined and predicted in terms of the properties of the electrons, protons, etc., of which its atoms are composed. And they would also argue that the properties of human society are similarly no more than the sums of the individual behaviors and tendencies of the individual humans of which that society is composed. Societies are aggressive because the individuals who compose them are aggressive, for instance. In formal language, reductionism is the claim that the compositional units of a whole are ontologically prior to the whole that the units comprise. That is, the units and their properties exist before the whole, and there is a chain of causation that runs from the units to the whole.

The second stance is related to the first; indeed, it is in some senses a special case of reductionism. It is that of biological determinism. Biological determinists ask, in essence, Why are individuals as they are? Why do they do what they do? And they answer that human lives and actions are inevitable consequences of the biochemical properties of the cells that make up the individual; and these characteristics are in turn uniquely determined by the constituents of the genes possessed by each individual. Ultimately, all human behavior—hence all human society—is governed by a chain of determinants that runs from the gene to the individual to the sum of the behaviors of all individuals. The determinists would have it, then, that human nature is fixed by our genes. The good society is either one in accord with a human nature to whose fundamental characteristics of inequality and competitiveness the ideology claims privileged access, or else it is an unattainable utopia because human nature is in unbreakable contradiction with an arbitrary notion of the good derived without reference to the facts of physical nature.

Science and Ideology

One of the issues with which we must come to grips is that, despite its frequent claim to be neutral and objective, science is not and cannot be above "mere" human politics. The complex interactions between the evolution of scientific theory and the evolution of social order means that very often the ways in which scientific research asks its questions of the human and natural worlds it proposes to explain are deeply colored by social, cultural, and political biases.

Critics of biological determinism have frequently drawn attention to the ideological role played by apparently scientific conclusions about the human condition that seem to flow from biological determinism. That, despite their pretensions, biological determinists are engaged in making political and moral statements about human society, and that their writings are seized upon as ideological legitimators, says nothing, in itself, about the scientific merits of their claims. Critics of biological determinism are often accused of merely disliking its political conclusions. We have no hesitation in agreeing that we do dislike these conclusions; we believe that it is possible to create a better society than the one we live in at present; that inequalities of wealth, power, and status are not "natural" but socially imposed obstructions to the building of a society in which the creative potential of all its citizens is employed for the benefit of all.

We view the links between values and knowledge as an integral part of doing science in this society at all, whereas determinists tend to deny that such links exist—or claim that if they do exist they are exceptional pathologies to be eliminated. To us such an assertion of the separation of fact from value, of practice from theory, science from society is itself part of the fragmentation of knowledge that reductionist thinking sustains and which has been part of the mythology of the last century of "scientific advance." However, the least of our tasks here is that of criticizing the social implications of biological determinism, as if the broad claims of biological determinism could be upheld. Rather, our major goal is to show that the world is not to be understood as biological determinism would have it be, and that, as a way of explaining the world, biological determinism is fundamentally flawed.

Note that we say "the world," for another misconception is that the criticism of biological determinism applies only to its conclusions about human societies, while what it says about nonhuman animals is more or less valid. Such a view is often expressed—for instance about E. O. Wilson's book *Sociobiology: The New Synthesis*. Its liberal critics claim that the problem with *Sociobiology* lies only in the first and last chapters, where the author discusses human sociobiology; what's in between is true. Not so, in our view: what biological determinism has to say about human society is more wrong than what it says about other aspects of biology because its simplifications and misstatements are the more gross. But this is not because it has developed a theory applicable only to nonhuman animals; the method and theory are fundamentally flawed whether applied to the United States or Britain today, or to a population of savanna-dwelling baboons or Siamese fighting fish.

Sociobiology and "Just So Stories"

It would be difficult for anyone to present the entire set of social phenomena that are said to be human nature. Indeed, there is disagreement even among sociobiologists on an appropriate list. Roughly, humans are seen as self-aggrandizing, selfish animals whose social organization, even in its cooperative aspects, is a consequence of natural selection for traits that maximize reproductive fitness. In particular, humans are characterized by territoriality, tribalism, indoctrinability, blind faith, xenophobia, and a variety of manifestations of aggression. Unselfish behavior is really a form of selfishness in which the individual is motivated by an expectation of reciprocal reward. Self-righteousness, gratitude, and sympathy are examples, while aggressively moralistic behavior is a way of keeping cheaters in line.

To universalize features of society through history and over cultures is not difficult. The very richness of the ethnographic record and the plasticity in its interpretations guarantee that large numbers of tribes said to display one phenomenon or another can be chosen as anecdotal cases. The amassing of supporting anecdotes is a standard method in works of advocacy. One element in the sociobiological argument is to reconstruct a plausible story for the origin of human social traits by natural selection. The general outline is to suppose that in the evolutionary past of the species there existed some genetic variation for a particular trait, but that the genotypes determining a particular form of behavior somehow left more offspring. As a consequence, these genotypes increased in the species and eventually came to characterize it. Imaginative stories have been told for ethics, religion, male domination, aggression, artistic ability, etc. All one need do is predicate a genetically determined contrast in the past and then use some imagination, in a Darwinian version of Kipling's *Just So Stories*.

"This is the Elephant's Child having his nose pulled by the Crocodile. He is very much surprised and astonished and hurt, and he is talking through his nose and saying, 'Led go! You are hurtig be!'" Drawing and caption by Rudyard Kipling from *Just So Stories*, 1928.

The combination of direct selection, kin selection, and reciprocal altruism provides the sociobiologist with a battery of speculative possibilities that guarantees an explanation for every observation. The system is unbeatable because it is insulated from any possibility of being contradicted by fact. If one is allowed to invent genes with arbitrarily complicated effects on phenotype and then to invent adaptive stories about the unrecoverable past of human history, all phenomena, real and imaginary, can be explained.

Abuses of Biological Determinism

The characteristic of science, as opposed to prerevolutionary natural philosophy, is that it is an activity of a special group of self-validating experts: scientists. The word "scientist" itself did not come into the language until 1840. The appeal to the "scientific" for legitimacy and to scientists as the ultimate authorities is quintessentially modern. The objectification of social relations which is embodied in science is translated into the objectivity, disinterestedness, and lack of passion of scientists (except their "passion for truth"). Since science is now the source of legitimacy for ideology, so scientists become the generators of the concrete form in which it enters public consciousness.

It is important to understand that biological determinism, even in its most gross and vicious forms, is not the product of a fringe of crackpots and vulgar popularizers, but of some of the core members of the university and scientific community. The Nobel Prize laureate, Konrad Lorenz, in a scientific paper on animal behavior in 1940 in Germany during the Nazi extermination campaign said, "The selection of toughness, heroism, social utility . . . must be accomplished by some human institutions if mankind, in default of selection factors, is not to be ruined by domestication induced degeneracy. The racial idea as the basis of the state has already accomplished much in this respect." He was only applying the view of the founder of eugenics, Sir Francis Galton, who sixty years before wondered that "there exists a sentiment, for the most part quite unreasonable, against the gradual extinction of an inferior race." What for Galton was a gradual process became rather more rapid in the hands of Lorenz's efficient friends.

In 1924 the Congress of the United States passed an immigration restriction act that weighted future immigration in the United States heavily against Eastern and Southern Europeans. Testimony before Congress by leaders of the American mental testing movement to the effect that Slavs, Jews, Italians, and others were mentally dull and that their dullness was racial, or at least constitutional, gave scientific legitimacy to the law that was constructed. Ten years later the same argument was the basis for the German racial and eugenic laws that began with the sterilization of the mentally and morally undesirable and ended in Auschwitz. The claims of biological determinists and eugenicists to scientific respectability were severely damaged in the gas chambers of the "Final Solution." At present, the National Front in Britain and the Nouvelle Droite in France argue that racism and anti-Semitism are natural and cannot be eliminated, citing as their authority E. O. Wilson of Harvard, who claims that territoriality, tribalism, and xenophobia are indeed part of the human genetic constitution, having been built into it by millions of years of evolution.

The Myth of Scientific Objectivity

Certainly the repugnant political consequences that have repeatedly flowed from determinist arguments are not criteria by which to judge their objective truth. We cannot derive "ought" from "is" or "is" from "ought." Biological determinists try to have it both ways. To legitimize theories they deny any connection to political events, giving the impression that the theories are the outcome of internal developments within a science that is insulated from social relations. They then become political actors, writing for newspapers and popular magazines, testifying before legislatures, appearing as celebrities on television to explicate the political and social consequences that must flow from their objective science. They change their personae from the scientific to the political and back again as the occasion demands, taking their legitimacy from science and their relevance from politics. They understand that, although there is no logical necessity connecting the truth of determinism to its political role, their own legitimacy as scientific authorities is dependent upon their appearance as politically disinterested parties. In this sense, biological determinists are victims of the very myth of the separation of science from social relations that they and their academic predecessors have perpetuated.

"Science" is sometimes taken to mean the body of scientists and the set of social institutions in which they participate, the journals, the books, the laboratories, the professional societies and academies through which individuals and their work are given currency and legitimacy. At other times "science" stands for the set of methods that are used by scientists as means for investigating the relations among things in the world, and the canons of evidence that are accepted as giving credibility to the conclusions of scientists. Yet a third meaning given to "science" is the body of facts, laws, theories, and relationships concerning real phenomena that the social institutions of "science," using the methods of "science," claim to be true.

It is extremely important for us to distinguish what the social institutions of science, using the methods of science, say about the world of phenomena from the actual world of phenomena itself. Just because those social institutions, using these methods, have so often said true things about the world, we are in danger of forgetting that sometimes the claims of those who speak in the name of "science" are rubbish.

Why, then, are they given such serious attention? It is because, in contemporary Western society, science as an institution has come to be accorded the authority that once went to the Church. When "science" speaks—or rather when its spokesmen (and they are generally men) speak in the name of science—let no dog bark. "Science" is the ultimate legitimator of bourgeois ideology. To oppose "science," to prefer values to facts, is to transgress not merely against a human law but against a law of nature.

Tests of Truth and Social Function

Let us be clear as to what it is we are maintaining about science and its claims: We are not arguing that to state the political philosophy or social position of the exponents of a particular scientific claim is enough to evaporate or invalidate that claim. Explaining its origins does not explain away the claim itself. (This is what philosophers call the "genetic fallacy.") We are arguing that there are two distinct questions to be asked of any description or explanation offered of the events, phenomena, and processes that occur in the world around us.

The first is about the internal logic and asks: Is the description accurate and the explanation true? That is, do they correspond to the reality of the phenomena, events, and processes in the real world? It is this type of question about the internal logic of science that most Western philosophers of science believe, or claim to believe, science to be all about. The model of scientific advance that most scientists are taught, and which is largely based on the writing of Karl Popper and his acolytes, sees science as progressing in this abstract way, by a continuous sequence of theory-making and testing, conjectures and refutations. In the more up-to-date Kuhnian version of the model, these conjectures and refutations of "normal" science are occasionally convulsed by periods of "revolutionary" science in which the entire framework ("paradigm") within which the conjectures and refutations are framed is shaken, like a kaleidoscope which relocates the same pieces of data into quite new patterns, even though the whole process of theory-making is believed to occur autonomously without reference to the social framework in which science is done.

But the second question to be asked of description or explanations is about the social matrix in which science is embedded—and it is a question of equal importance. The insight into the theories of scientific growth hinted at in the nineteenth century by Marx and Engels, developed by a generation of Marxist scholars in the 1930s, and now reflected, refracted, and plagiarized by a host of sociologists, is that scientific growth does not proceed in a vacuum. The questions asked by scientists, the types of explanation accepted as appropriate, the paradigms framed, and the criteria for weighing evidence are all historically relative. They do not proceed from some abstract contemplation of the natural world as if scientists were programmable computers who neither made love, ate, defecated, had enemies, nor expressed political views. In particular, scientists are seen as individuals confronting an external and objective nature, wrestling with nature to extract its secrets, rather than as people with particular relations to each other, to the state, to their patrons, and to the owners of wealth and production. Thus, scientists are defined as those who do science, rather than science being defined as what scientists do.

But scientists have done more than simply participate in the general objectification of society. They have raised that objectification to the status of an absolute good called "scientific objectivity." Just as the objectification of society in general unleashed the immense productive forces of capitalism, so scientific objectivity in particular was a progressive step toward gaining real knowledge about the world. Such objectivity, as we all recognize, has been responsible for an immense increase in the power to manipulate the world for human purposes. But the emphasis on objectivity has masked the true social relations of scientists with each other and with the rest of society. By denying these relations, scientists make themselves vulnerable to a loss of credibility and legitimacy when the mask slips and the social reality is revealed.

Thus, at any historical moment, what pass as acceptable scientific explanations have both social determinants and social functions. The progress of science is the product of a continuous tension between the internal logic of a method of acquiring knowledge that professes correspondence with and truth about the real material world, and the external logic of these social determinants and functions. Those conservative philosophers who deny the latter, and some more currently fashionable sociologists who wish to dissolve away the former entirely, alike fail to understand the power and the role of this tension, which forms the essential dynamic of science whose ultimate tests are always twofold: tests of truth and of social function.

A Dialectical Alternative

We do not offer in this book a blueprint or a catalogue of certainties; our task, as we see it, is to point the way toward an integrated understanding of the relationship between the biological and the social. We describe such an understanding as dialectical, in contrast to reductionist. *(Richard Lewontin is coauthor with Richard Levins of* The Dialectical Biologist, *1985.)*

Reductionist explanation attempts to derive the properties of wholes from intrinsic properties of parts, properties that exist apart from and before the parts are assembled into complex structures. It is characteristic of reductionism that it assigns relative weights to different partial causes and attempts to assess the importance of each cause by holding all others constant while varying a single factor. Dialectical explanations, on the contrary, do not abstract properties of parts in isolation from their associations in wholes but see the properties of parts as arising out of their associations. That is, according to the dialectical view, the properties of parts and wholes codetermine each other. The properties of individual human beings do not exist in isolation but arise as a consequence of social life, yet the nature of that social life is a consequence of our being human and not, say, plants. It follows, then, that dialectical explanation contrasts with cultural or dualistic modes of explanation that separate the world into different types of phenomena—culture and biology, mind and body—which are to be explained in quite different and nonoverlapping ways.

Dialectical explanations attempt to provide a coherent, unitary, but nonreductionist account of the material universe. For dialectics the universe is unitary but always in change; the phenomena we can see at any instant are parts of processes, processes with histories and futures whose paths are not uniquely determined by their constituent units. Wholes are composed of units whose properties may be described, but the interaction of these units in the construction of the wholes generates complexities that result in products qualitatively different from the component parts. Such a world view abolishes the antitheses of reductionism and dualism; of nature/nurture or of heredity/environment; of a world in stasis whose components interact in fixed and limited ways, indeed in which change is possible only along fixed and previously definable pathways.

Let us take just one example here, that of the relationship of the organism to its environment. Biological determinism sees organisms, human or nonhuman, as adapted by evolutionary processes to their environment, that is, fitted by the processes of genetic reshuffling, mutation, and natural selection to maximize their reproductive success in the environment in which they are born and develop. Further, it sees the undoubted plasticity of organisms—especially humans—as they develop as a series of modifications imposed upon an essentially passive, recipient object by the buffeting of "the environment" to which it is exposed and to which it must adapt or perish. Against this we counterpose a view not of organism and environment insulated from one another or unidirectionally affected, but of a constant and active interpenetration of the organism with its environment. Organisms do not merely receive a given environment but actively seek alternatives or change what they find.

Put a drop of sugar solution onto a dish containing bacteria and they will actively move toward the sugar till they arrive at the site of optimum concentration, thus changing a low-sugar for a high-sugar environment. They will then actively work on the sugar molecules, changing them into other constituents, some of which they absorb, others of which they put out into the environment, thereby modifying it, often in such a way that it becomes, for example, more acid. When this happens, the bacteria move away from the highly acid region to regions of lower acidity. Here, in miniature, we see the case of an organism "choosing" a preferred environment, actively working on it and so changing it, and then "choosing" an alternative.

Or consider a bird building a nest. Straw is not part of the bird's environment unless it actively seeks it out so as to construct its nest; in doing so it changes its environment, and indeed the environment of other organisms as well. The "environment" itself is under constant modification by the activity of all the organisms within it. And to any organism, all others form part of its "environment"—predators, prey, and those that merely change the landscape it resides in.

Society: More Than the Sum of Its Parts

One of the chief claims of sociobiology is that society is constrained by individual properties that are translated as prohibitions on society. Yet the most striking feature of social life is that it so often is the negation of individual limitations. Indeed, that negation is the force that keeps societies together. People can do in concert what they cannot do separately. Nor is this property simply the result of the summation of individual forces, as when ten people can lift a weight that one person alone cannot. On the contrary, totally new properties arise from social interaction.

None of us can fly by flapping our arms either singly or in a crowd. Yet we do fly as a result of technology, airplanes, pilots, airlines, ground crew, all de novo products of social activity, qualitatively different from our individual acts. Moreover, it is not society that flies, but individuals.

The memories of individuals are limited, and if all the historians in the world were set to the task, they could not learn by rote even a tiny fraction of the factual material (the census figures, for instance) they use in their profession. Yet they can recall these facts, as individuals, by going to the library and reading books, a qualitatively new product of social activity. Once again, individuals acquire new properties from society.

At the same time, society is obviously made up of individuals. Society is not, in a metaphor that has persisted in various forms through many centuries, itself an organism. It is not a Platonic form that has an independent existence above and outside of individual people. It is their creation. It is, as Marx said, "men that change circumstances." While Newtonian mechanics would have come into existence even if Isaac Newton had died in his crib, it was, in fact, a product of individual thought.

Society does not think; only individuals think. Thus, the relation between individual and society, like the relations between organism and

Terraced rice agriculture in Bali.

environment, is a dialectical one. It is not only that society is the environment of the individual and therefore perturbs and is perturbed by the individual. Society is also hierarchically related to individuals. As a collection of individual lives, it possesses some structural properties, just as all collections have properties that are not properties of the individuals that make them up, while at the same time lacking certain properties of the individuals. Only an individual can think, but only a society can have a class structure. At the same time, what makes the relation between society and the individual dialectical is that individuals acquire from the society produced by them individual properties, like flying, that they did not possess in isolation. It is not just that wholes are more than the sum of their parts; it is that parts become qualitatively new by being parts of the whole.

What characterizes human development and actions is that they are the consequence of an immense array of interacting and intersecting causes. Our actions are not at random or independent with respect to the totality of those causes as an intersecting system, for we are material beings in a causal world. But to the extent that they are free, our actions are independent of any one or even a small subset of those multiple paths of causation: that is the precise meaning of freedom in a causal world. When, on the contrary, our actions are predominantly constrained by a single cause, like the train on the track, the prisoner in his cell, the poor person in her poverty, we are no longer free.

For biological determinists we are unfree because our lives are strongly constrained by a relatively small number of internal causes, the genes for specific behaviors or for predisposition to these behaviors. But this misses the essence of the difference between human biology and that of other organisms. Our brains, hands, and tongues have made us independent of many single major features of the external world. Our biology has made us into creatures who are constantly re-creating our own psychic and material environments, and whose individual lives are the outcomes of an extraordinary multiplicity of intersecting causal pathways. Thus, it is our biology that make us free.

Closing Arguments

Not in Our Genes *is only one of many books that criticize sociobiology.* Vaulting Ambition *by Philip Kitcher (a professor of philosophy at the University of California, San Diego) is another widely quoted text. In this 1987 book (reprinted by permission of The MIT Press), Kitcher makes some of the same points as those presented above. But his main comment is this:*

Philip Kitcher

The dispute about human sociobiology is a dispute about evidence. Friends of sociobiology see the "new synthesis" as an exciting piece of science, resting soundly on evidence and promising a wealth of new insights, including some that are relevant to human needs. To critical eyes, however, the same body of doctrine seems a mass of unfounded speculation, mischievous in covering socially harmful suggestions with the trappings and authority of science.

Everybody ought to agree that, given sufficient evidence for some hypothesis about humans, we should accept that hypothesis whatever its political implications. But the question of what counts as sufficient evidence is not independent of the political consequences. If the costs of being wrong are sufficiently high, then it is reasonable and responsible to ask for more evidence than is demanded in situations where mistakes are relatively innocuous.

In the free-for-all of scientific research, ideas are often tossed out, tentatively accepted, and only subsequently subjected to genuinely rigorous tests. Arguably, the practice of bold overgeneralization contributes to the efficient working of science as a community enterprise: hypotheses for which there is "some evidence" or, perhaps, "reasonably good evidence" become part of the public fund of ideas, are integrated with other hypotheses, criticized, refined, and sometimes discarded. Yet when the hypotheses in question bear on human concerns, the exchange cannot be quite so cavalier.

If a single scientist, or even the whole community of scientists, comes to adopt an incorrect view of the origins of a distant galaxy, an inadequate model of foraging behavior in ants, or a crazy explanation of the extinction of the dinosaurs, then the mistake will not prove tragic. By contrast, if we are wrong about the bases of human social behavior, if we abandon the goal of a fair distribution of the benefits and burdens of society because we accept faulty hypotheses about ourselves and our evolutionary history, then the consequences of a scientific mistake may be grave indeed.

The moral is transparent. When scientific claims bear on matters of social policy, the standards of evidence and of self-criticism must be extremely high.

Ashley Montagu, another critic of sociobiology, cautions readers to examine the underlying biases of both supporters and critics. In his 1980 book, Sociobiology Examined *(reprinted here by permission of Oxford University Press), he says,*

Ashley Montagu

When in 1975, Edward O. Wilson's book *Sociobiology: The New Synthesis* appeared, it was greeted with the kind of attention that few books of its kind have received in our time. Except for a relatively few dissenters it was widely and favorably reviewed in the scientific, literary, and lay press. The book created something of a sensation principally because of its claim to have laid the foundations for a new science, one that would for the first time provide a firm biological basis for the understanding of the refractory human social behavior with which social scientists have ineffectually attempted to grapple for too long a time.

Having written a quite admirable 546 pages of beautifully illustrated text setting out the principles of the new discipline of sociobiology—which Wilson defines as the systematic study of the biological basis of all forms of social behavior, including sexual and parental behavior, in all kinds of organisms, including man—Wilson devotes his concluding chapter to "Man: From Sociobiology to Sociology." It is this twenty-seventh chapter which has given rise to much criticism and a continuing debate which promises to be as lively as any in recent decades.

Unfortunately some of the discussion has been marred by a bias and an occasional intemperateness which has only served to lose sympathy for the guilty ones. Wilson has withstood these assaults upon his integrity with civility and the appropriate sense of humor. Truth will not be advanced by heat so much as it will by light. Nor will it be advanced by the kind of loose thinking and blindered ignorance that has characterized some of the principal participants in the debate.

It is a matter of general experience that where nature-nurture issues are concerned, emotion only too often enters in. It is strange to see the "mature" "dispassionate scientist" behaving in this way only because the "mature" "dispassionate scientist" is a myth. Scientists are human beings with their full complement of emotions and prejudices, and their emotions and prejudices often influence the way they do their science.

This was first clearly brought out in a study by Professor Nicholas Pastore, *The Nature-Nurture Controversy*, published in 1949. In this study Professor Pastore showed that the scientist's political beliefs were highly correlated with what he believed about the roles played by nature and nurture in the development of the person. Those holding conservative political views strongly tended to believe in the power of genes over environment. Those subscribing to more liberal views tended to believe in the power of environment over genes. One distinguished scientist (who happened to be a teacher of mine) when young was a socialist and environmentalist but toward middle age he became politically conservative and a firm believer in the supremacy of genes!

Today we see a similar division between those who are politically left and those who are politically right of center. I mention these matters because I think they are important, and because in spite of all attempts to disavow the role that political views and private emotions have played in the presentation of sociobiological ideas and their criticism it is clear that such biases have frequently been involved. It is well to remember this in reading any discussion of sociobiology—for or against.

Finally, the evolution theorist Stephen Jay Gould too has been a vocal critic of sociobiology:

Stephen Jay Gould

When sociobiology is injudicious and trades in speculative genetic arguments about specific human behaviors, it speaks nonsense. When it is judicious and implicates genetics only in setting the capacity for broad spectra of culturally conditioned behaviors, then it is not very enlightening. To me, such an irresolvable dilemma only indicates that this latest attempt to reduce the human sciences will have very limited utility.

After all this criticism, fairness demands that the proponents of sociobiology be given a brief chance to refresh our awareness of their views. I will therefore close this chapter with Richard D. Alexander's statement from his 1987 book, The Biology of Moral Systems.

Richard D. Alexander

To say we are evolved to serve the interests of our genes in no way suggests that we are obliged to serve them. Evolution is surely most deterministic for those still unaware of it.

VI *Selfish Genes*

A hen is only an egg's way of making another egg.

—Samuel Butler

We are survival machines—robot vehicles blindly programmed to preserve the selfish molecules known as genes.

—Richard Dawkins

Edward O. Wilson and William Hamilton made their breakthroughs in biology by taking a gene's-eye view of animal (and human) behavior. For more than a century, scientists had puzzled over the presence of altruism throughout the animal kingdom. Wilson and Hamilton found that if we look beyond the bounds of the organism, purposeful actions hazardous or lethal for an individual (like the stinging honeybee) are actually beneficial for copies of the individual's genes.

Richard Dawkins (an ethologist at Oxford University) has taken this gene's-eye view of life one step further in his 1976 book, The Selfish Gene. *Now that sociobiologists have made sense of seemingly nonsensical behavior, Dawkins extends their ideas in an attempt to grasp the very meaning and purpose of life. Evolution theorist John Maynard Smith finds* The Selfish Gene *unusual in several ways.*

A New Worldview

John Maynard Smith

Although written as a popular account, it made an original contribution to biology. Further, the contribution itself was of an unusual kind. *The Selfish Gene* reports no new facts. Nor does it contain any new mathematical models—indeed it contains no mathematics at all. What it does offer is a new worldview.

Although the book has been widely read and enjoyed, it has also aroused strong hostility. Much of this hostility arises, I believe, from misunderstanding, or rather, from several misunderstandings. Of these, the most fundamental is a failure to understand what the book is about. It is a book about the evolutionary process—it is not about morals, or about politics, or about the human sciences. If you are not interested in how evolution came about, and cannot conceive how anyone could seriously be concerned about anything other than human affairs, then do not read it: it will only make you needlessly angry.

What is this "new worldview" that prompts such passion? Richard Dawkins states his central thesis thus:

Richard Dawkins

We are survival machines—robot vehicles blindly programmed to preserve the selfish molecules known as genes. This is a truth which still fills me with astonishment. Though I have known it for years, I never seem to get fully used to it. One of my hopes is that I may have some success in astonishing others.

Darwin's Theory of Evolution

Excerpts from The Selfish Gene *by Richard Dawkins (copyright 1976 by Oxford University Press; new edition copyright 1989 by Richard Dawkins) are here reprinted by permission of the author and Oxford University Press. All section headings have been added by the editor.*

Richard Dawkins

Intelligent life on a planet comes of age when it first works out the reason for its own existence. If superior creatures from space ever visit earth, the first question they will ask, in order to assess the level of our civilization, is: "Have they discovered evolution yet?" Living organisms had existed on earth, without ever knowing why, for over three thousand million years before the truth finally dawned on one of them. His name was Charles Darwin. To be fair, others had had inklings of the truth, but it was Darwin who first put together a coherent and tenable account of why we exist.

We no longer have to resort to superstition when faced with the deep problems: Is there a meaning to life? What are we for? What is man? After posing the last of these questions, the eminent zoologist George Gaylord Simpson put it thus: "The point I want to make now is that all attempts to answer that question before 1859 are worthless and that we will be better off if we ignore them completely."

Today the theory of evolution is about as much open to doubt as the theory that the earth goes round the sun, but the full implications of Darwin's revolution have yet to be widely realized. Zoology is still a minority subject in universities, and even those who choose to study it often make their decision without appreciating its profound philosophical significance. Philosophy and the subjects known as "humanities" are still taught almost as if Darwin had never lived. No doubt this will change in time. In any case, this book is not intended as a general advocacy of Darwinism. Instead, it will explore the consequences of the evolution theory for a particular issue. My purpose is to examine the biology of selfishness and altruism.

Apart from its academic interest, the human importance of this subject is obvious. It touches every aspect of our social lives, our loving and hating, fighting and cooperating, giving and stealing, our greed and our generosity. These are claims which could have been made for Lorenz's *On Aggression*, Ardrey's *The Social Contract*, and Eibl-Eibesfeldt's *Love and Hate*. The trouble with these books is that their authors got it totally and utterly wrong. They got it wrong because they misunderstood how evolution works.

They made the erroneous assumption that the important thing in evolution is the good of the species (or the group) rather than the good of the individual (or gene). It is ironic that Ashley Montagu should criticize Lorenz as a "direct descendant of the 'nature red in tooth and claw' thinkers of the nineteenth century." As I understand Lorenz's view of evolution, he would be very much at one with Montagu in rejecting the implications of Tennyson's famous phrase. Unlike both of them, I think "nature red in tooth and claw" sums up our modern understanding of natural selection admirably.

The Law of Ruthless Selfishness

Before beginning on my argument itself, I want to explain briefly what sort of an argument it is, and what sort of an argument it is not. If we were told that a man had lived a long and prosperous life in the world of Chicago gangsters, we would be entitled to make some guesses as to the sort of man he was. We might expect that he would have qualities such as toughness, a quick trigger finger, and the ability to attract loyal friends. These would not be infallible deductions, but you can make some inferences about a man's character if you know something about the conditions in which he has survived and prospered.

The argument of this book is that we, and all other animals, are machines created by our genes. Like successful Chicago gangsters, our genes have survived, in some cases for millions of years, in a highly competitive world. This entitles us to expect certain qualities in our genes. I shall argue that a predominant quality to be expected in a successful gene is ruthless selfishness. This gene selfishness will usually give rise to selfishness in individual behavior. However, there are special circumstances in which a gene can achieve its own selfish goals best by fostering a limited form of altruism at the level of individual animals. "Special" and "limited" are important words in the last sentence. Much as we might wish to believe otherwise, universal love and the welfare of the species as a whole are concepts which simply do not make evolutionary sense.

This brings me to the first point I want to make about what this book is not. I am not advocating a morality based on evolution. I am saying how things have evolved. I am not saying how we humans morally ought to behave. I stress this, because I know I am in danger of being misunderstood by those people, all too numerous, who cannot distinguish between a statement of belief in what is the case from an advocacy of what ought to be the case. My own feeling is that a human society based simply on the gene's law of universal ruthless selfishness would be a very nasty society in which to live. But unfortunately, however much we may deplore something, it does not stop it being true.

This book is mainly intended to be interesting, but if you would extract a moral from it, read it as a warning. Be warned that if you wish, as I do, to build a society in which individuals cooperate generously and unselfishly towards a common good, you can expect little help from biological nature. Let us try to teach generosity and altruism, because we are born selfish. Let us understand what our own selfish genes are up to, because we may then at least have the chance to upset their designs, something which no other species has ever aspired to.

'Primeval Soup' and the Origin of Life

The argument takes time to develop, and we must begin at the beginning, with the very origin of life itself. The account of the origin of life which I shall give is necessarily speculative; by definition, nobody was around to see what happened. There are a number of rival theories, but they all have certain features in common. The simplified account I shall give is probably not too far from the truth.

We do not know what chemical raw materials were abundant on earth before the coming of life, but among the plausible possibilities are water, carbon dioxide, methane, and ammonia: all simple compounds known to be present on at least some of the other planets in our solar system. Chemists have tried to imitate the chemical conditions of the young earth. They have put these simple substances in a flask and supplied a source of energy such as ultraviolet light or electric sparks—artificial simulation of primordial lightning. After a few weeks of this, something interesting is usually found inside the flask: a weak brown soup containing a large number of molecules more complex than the ones originally put in. In particular, amino acids have been found—the building blocks of proteins, one of the two great classes of biological molecules.

Before these experiments were done, naturally-occurring amino acids would have been thought of as diagnostic of the presence of life. If they had been detected on, say Mars, life on that planet would have seemed a near certainty. Now, however, their existence need imply only the presence of a few simple gases in the atmosphere and some volcanoes, sunlight, or thundery weather. More recently, laboratory simulations of the chemical conditions of earth before the coming of life have yielded organic substances called purines and pyrimidines. These are building blocks of the genetic molecule, DNA itself.

Processes analogous to these must have given rise to the 'primeval soup' which biologists and chemists believe constituted the seas some three to four thousand million years ago. The organic substances became locally concentrated, perhaps in drying scum round the shores, or in tiny suspended droplets. Under the further influence of energy such as ultraviolet light from the sun, they combined into larger molecules. Nowadays large organic molecules would not last long enough to be noticed: they would be quickly absorbed and broken down by bacteria or other living creatures. But bacteria and the rest of us are late-comers, and in those days large organic molecules could drift unmolested through the thickening broth.

Rise of the Replicators

At some point a particularly remarkable molecule was formed by accident. We will call it the replicator. It may not necessarily have been the biggest or the most complex molecule around, but it had the extraordinary property of being able to create copies of itself. This may seem a very unlikely sort of accident to happen. So it was. It was exceedingly improbable. In the lifetime of a man, things which are that improbable can be treated for all practical purposes as impossible. That is why you will never win a big prize on the football pools. But in our human estimates of what is probable and what is not, we are not used to dealing in hundreds of millions of years. If you filled in pools coupons every week for a hundred million years you would very likely win several jackpots.

Actually, a molecule which makes copies of itself is not as difficult to imagine as it seems at first, and it only had to arise once. Think of the replicator as a mould or template. Imagine it as a large molecule consisting of a complex chain of various sorts of building block molecules. The small building blocks were abundantly available in the soup surrounding the replicator. Now suppose that each building block has an affinity for its own kind. Then whenever a building block from out in the soup lands up next to a part of the replicator for which it has an affinity, it will tend to stick there. The building blocks which attach themselves in this way will automatically be arranged in a sequence which mimics that of the replicator itself. It is easy then to think of them joining up to form a stable chain just as in the formation of the original replicator. This process could continue as a progressive stacking up, layer upon layer. This is how crystals are formed. On the other hand, the two chains might split apart, in which case we have two replicators, each of which can go on to make further copies.

A more complex possibility is that each building block has affinity not for its own kind, but reciprocally for one particular other kind. Then the replicator would act as a template not for an identical copy, but for a kind of 'negative,' which would in its turn re-make an exact copy of the original positive. For our purposes it does not matter whether the original replication process was positive-negative or positive-positive, though it is worth remarking that the modern equivalents of the first replicator, the DNA molecules, use positive-negative replication.

What does matter is that suddenly a new kind of stability came into the world. Previously it is probable that no particular kind of complex molecule was very abundant in the soup, because each was dependent on building blocks happening to fall by luck into a particular stable configuration. As soon as the replicator was born it must have spread its copies rapidly throughout the seas, until the smaller building block molecules became a scarce resource, and other larger molecules were formed more and more rarely.

Longevity, Fecundity, and Copying-Fidelity

So we seem to arrive at a large population of identical replicas. But now we must mention an important property of any copying process: it is not perfect. Mistakes will happen. I hope there are no misprints in this book, but if you look carefully you may find one or two. They will probably not seriously distort the meaning of the sentences, because they will be 'first generation' errors. But imagine the days before printing, when books such as the Gospels were copied by hand. All scribes, however careful, are bound to make a few errors, and some are not above a little willful 'improvement.' If they all copied from a single master original, meaning would not be greatly perverted. But let copies be made from other copies, which in their turn were made from other copies, and errors will start to become cumulative and serious.

We tend to regard erratic copying as a bad thing, and in the case of human documents it is hard to think of examples where errors can be described as improvements. I suppose the scholars of the Septuagent could at least be said to have started something big when they mistranslated the Hebrew word for 'young woman' into the Greek word for 'virgin,' coming up with the prophecy: "Behold a virgin shall conceive and bear a son." Anyway, as we shall see, erratic copying in biological replicators can in a real sense give rise to improvement, and it was essential for the progressive evolution of life that some errors were made.

We do not know how accurately the original replicator molecules made their copies. Their modern descendants, the DNA molecules, are astonishingly faithful compared with the most high-fidelity human copying process, but even they occasionally make mistakes, and it is ultimately these mistakes which make evolution possible. Probably the original replicators were far more erratic, but in any case we may be sure that mistakes were made, and these mistakes were cumulative. As mis-copyings were made and propagated, the primeval soup became filled by a population not of identical replicas, but of several varieties of replicating molecules, all 'descended' from the same ancestor. Would some varieties have been more numerous than others? Almost certainly yes. Some varieties would have been inherently more stable than others. Certain molecules, once formed, would be less likely than others to break up again. These types would become relatively numerous in the soup, not only as a direct logical consequence of their longevity, but also because they would have a long time available for making copies of themselves. Replicators of high longevity would therefore tend to become more numerous and, other things being equal, there would have been an evolutionary trend towards greater longevity in the population of molecules.

But other things were probably not equal, and another property of a replicator variety which must have had even more importance in spreading it through the population was speed of replication or fecundity. If replicator molecules of type *A* make copies of themselves on average once a week while those of type *B* make copies of themselves once an hour, it is not difficult to see that pretty soon type *A* molecules are going to be far outnumbered, even if they 'live' much longer than *B* molecules. There would therefore probably have been an evolutionary trend towards higher fecundity of molecules in the soup.

A third characteristic of replicator molecules which would have been positively selected is accuracy of replication. If molecules of type *X* and type *Y* last the same length of time and replicate at the same rate, but *X* makes a mistake on average every tenth replication while *Y* makes a mistake only every hundredth replication, *Y* will obviously become more numerous. If you already know something about evolution, you may find something slightly paradoxical about the last point. Can we reconcile the idea that copying errors are an essential prerequisite for evolution to occur, with the statement that natural selection favours high copying-fidelity? The answer is that although evolution may seem, in some vague sense, a 'good thing,' especially since we are the product of it, nothing actually 'wants' to evolve. Evolution is something that happens, willy-nilly, in spite of all the efforts of the replicators (and nowadays of the genes) to prevent it happening.

To return to the primeval soup, it must have become populated by stable varieties of molecule; stable in that either the individual molecules lasted a long time, or they replicated rapidly, or they replicated accurately. Evolutionary trends toward these three kinds of stability took place in the following sense: if you had sampled the soup at two different times, the later sample would have contained a higher proportion of varieties with high longevity/fecundity/copying-fidelity. This is essentially what a biologist means by evolution when he is speaking of living creatures, and the mechanism is the same—natural selection.

Resource Scarcity Begets Competition

Should we then call the original replicator molecules 'living'? Who cares? I might say to you "Darwin was the greatest man who has ever lived," and you might say "No, Newton was," but I hope we would not prolong the argument. The point is that no conclusion of substance would be affected whichever way our argument was resolved. The facts of the lives and achievements of Newton and Darwin remain totally unchanged whether we label them 'great' or not.

Similarly, the story of the replicator molecules probably happened something like the way I am telling it, regardless of whether we choose to call them 'living.' Human suffering has been caused because too many of us cannot grasp that words are only tools for our use, and that the mere presence in the dictionary of a word like 'living' does not mean it necessarily has to refer to something definite in the real world. Whether we call the replicators living or not, they were the ancestors of life; they were our founding fathers.

The next important link in the argument, one which Darwin himself laid stress on (although he was talking about animals and plants, not molecules) is competition. The primeval soup was not capable of supporting an infinite number of replicator molecules. For one thing, the earth's size is finite, but other limiting factors must also have been important. In our picture of the replicator acting as a template or mould, we supposed it to be bathed in a soup rich in the small building block molecules necessary to make copies. But when the replicators became numerous, building blocks must have been used up at such a rate that they became a scarce and precious resource. Different varieties or strains of replicator must have competed for them.

We have considered the factors which would have increased the numbers of favoured kinds of replicator. We can now see that less-favoured varieties must actually have become less numerous because of competition, and ultimately many of their lines must have gone extinct. There was a struggle for existence among replicator varieties. They did not know they were struggling, or worry about it; the struggle was conducted without any hard feelings, indeed without feelings of any kind. But they were struggling, in the sense that any miscopying which resulted in a new higher level of stability, or a new way of reducing the stability of rivals, was automatically preserved and multiplied.

Competition Begets Survival Machines

The process of improvement was cumulative. Ways of increasing stability and decreasing rivals' stability became more elaborate and more efficient. Some of them may even have 'discovered' how to break up molecules of rival varieties chemically, and to use the building blocks so released for making their own copies. These proto-carnivores simultaneously obtained food and removed competing rivals. Other replicators perhaps discovered how to protect themselves, either chemically, or by building a physical wall of protein around themselves. This may have been how the first living cells appeared. Replicators began not merely to exist, but to construct for themselves containers, vehicles for their continued existence. The replicators which survived were the ones which built survival machines for themselves to live in. The first survival machines probably consisted of nothing more than a protective coat. But making a living got steadily harder as new rivals arose with better and more effective survival machines. Survival machines got bigger and more elaborate, and the process was cumulative and progressive.

Was there to be any end to the gradual improvement in the techniques and artifices used by the replicators to ensure their own continuance in the world? There would be plenty of time for improvement. What weird engines of self-preservation would the millennia bring forth? Four thousand million years on, what was to be the fate of the ancient replicators?

They did not die out, for they are past masters of the survival arts. But do not look for them floating loose in the sea; they gave up that cavalier freedom long ago. Now they swarm in huge colonies, safe inside gigantic lumbering robots, sealed off from the outside world, communicating with it by tortuous indirect routes, manipulating it by remote control. They are in you and in me; they created us, body and mind; and their preservation is the ultimate rationale for our existence. They have come a long way, those replicators. Now they go by the name of genes, and we are their survival machines.

The gramineous bicycle . . . (Max Ernst, 1920/1921)

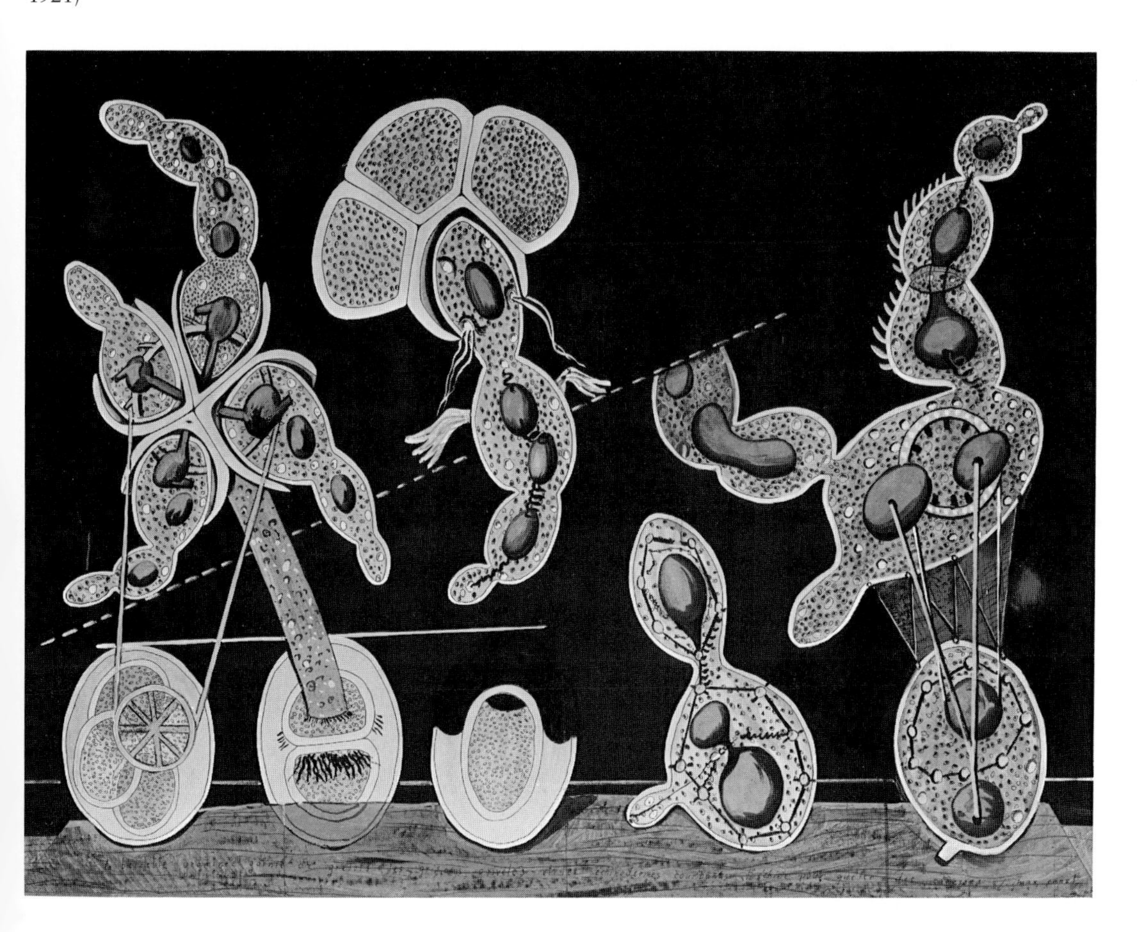

Today's Replicators: DNA

We are survival machines, but "we" does not mean just people. It embraces all animals, plants, bacteria, and viruses. The total number of survival machines on earth is very difficult to count and even the total number of species is unknown. Taking just the insects alone, the number of living species has been estimated at around three million, and the number of individual insects may be a million million million.

Different sorts of survival machine appear very varied on the outside and in their internal organs. An octopus is nothing like a mouse, and both are quite different from an oak tree. Yet in their fundamental chemistry they are rather uniform, and, in particular, the replicators which they bear, the genes, are basically the same kind of molecule in all of us—from bacteria to elephants.

We are all survival machines for the same kind of replicator—molecules called DNA—but there are many different ways of making a living in the world, and the replicators have built a vast range of machines to exploit them. A monkey is a machine which preserves genes up trees, a fish is a machine which preserve genes in the water; there is even a small worm which preserves genes in German beer mats. DNA works in mysterious ways.

For simplicity, I have given the impression that modern genes, made of DNA, are much the same as the first replicators in the primeval soup. It does not matter for the argument, but this may not really be true. The original replicators may have been a related kind of molecule to DNA, or they may have been totally different. In the latter case we might say that their survival machines must have been seized at a later stage by DNA. If so, the original replicators were utterly destroyed, for no trace of them remains in modern survival machines. Along these lines, A. G. Cairns-Smith has made the intriguing suggestion that our ancestors, the first replicators, may have been not organic molecules at all, but inorganic crystals—mineral, little bits of clay. Usurper or not, DNA is in undisputed charge today, unless, as I tentatively suggest in the final chapter, a new seizure of power is now just beginning.

A DNA molecule is a long chain of building blocks, small molecules called nucleotides. Just as protein molecules are chains of amino acids, so DNA molecules are chains of nucleotides. A DNA molecule is too small to be seen, but its exact shape has been ingeniously worked out by indirect means. It consists of a pair of nucleotide chains twisted together in an elegant spiral; the "double helix," the "immortal coil." *(In 1989 new technology enabled scientists actually to 'see' a DNA molecule for the first time, which confirmed once and for all Watson and Crick's 1953 hypothesis.)*

The nucleotide building blocks come in only four different kinds, whose names may be shortened to A, T, C, and G. These are the same in all animals and plants. What differs is the order in which they are strung together. A G building block from a man is identical in every particular to a G building block from a snail. But the sequence of building blocks in a man is not only different from that in a snail. It is also different—though less so—from the sequence in every other man (except in the special case of identical twins).

Our DNA lives inside our bodies. It is not concentrated in a particular part of the body, but is distributed among the cells. There are about a thousand million million cells making up an average human body, and, with some exceptions which we can ignore, every one of these cells contains a complete copy of that body's DNA. This DNA can be regarded as a set of instructions for how to make a body, written in the A, T, C, G alphabet of the nucleotides.

It is as though, in every room of a gigantic building, there was a bookcase containing the architect's plans for the entire building. The 'bookcase' in a cell is called the nucleus. The architect's plans run to 46 volumes in man—the number is different in other species. The 'volumes' are called chromosomes. They are visible under a microscope as long threads, and the genes are strung out along them in order. It is not easy, indeed it may not even be meaningful, to decide where one gene ends and the next one begins.

DNA molecules do two important things. Firstly they replicate, that is to say they make copies of themselves. This has gone on non-stop ever since the beginning of life, and the DNA molecules are now very good at it indeed. It is one thing to speak of the duplication of DNA. But if the DNA is really a set of plans for building a body, how are the plans put into practice? How are they translated into the fabric of the body? This brings me to the second important thing DNA does. It indirectly supervises the manufacture of a different kind of molecule—protein. The coded message of the DNA, written in the four-letter nucleotide alphabet, is translated in a simple mechanical way into another alphabet. This is the alphabet of amino acids which spells out protein molecules.

Making proteins may seem a far cry from making a body, but it is the first small step in that direction. Proteins not only constitute much of the physical fabric of the body; they also exert sensitive control over all the chemical processes inside the cell, selectively turning them on and off at precise times and in precise places. Exactly how this eventually leads to the development of a baby is a story which it will take decades, perhaps centuries, for embryologists to work out. But it is a fact that it does. Genes do indirectly control the manufacture of bodies, and the influence is strictly one way: acquired characteristics are not inherited. No matter how much knowledge and wisdom you acquire during your life, not one jot will be passed on to your children by genetic means. Each new generation starts from scratch. A body is the genes' way of preserving the genes unaltered.

The evolutionary importance of the fact that genes control embryonic development is this: it means that genes are at least partly responsible for their own survival in the future, because their survival depends on the efficiency of the bodies in which they live and which they helped to build. Once upon a time, natural selection consisted of the differential survival of replicators floating free in the primeval soup. Now, natural selection favours replicators which are good at building survival machines, genes which are skilled in the art of controlling embryonic development. In this, the replicators are no more conscious or purposeful than they ever were. The same old processes of automatic selection between rival molecules by reason of their longevity, fecundity, and copying-fidelity, still go on as blindly and as inevitably as they did in the far-off days. Genes have no foresight. They do not plan ahead. Genes just *are*, some genes more so than others, and that is all there is to it. But the qualities which determine a gene's longevity and fecundity are not so simple as they were. Not by a long way.

In recent years—the last six hundred million or so—the replicators have achieved notable triumphs of survival-machine technology such as the muscle, the heart, and the eye (evolved several times independently). Before that, they radically altered fundamental features of their way of life as replicators, which must be understood if we are to proceed with the argument.

A Short Course in Genetics

The first thing to grasp about a modern replicator is that it is highly gregarious. A survival machine is a vehicle containing not just one gene but many thousands. The manufacture of a body is a cooperative venture of such intricacy that it is almost impossible to disentangle the contribution of one gene from that of another. A given gene will have many different effects on quite different parts of the body. A given part of the body will be influenced by many genes, and the effect of any one gene depends on interaction with many others. Some genes act as master genes controlling the operation of a cluster of other genes. This intricate inter-dependence of genes may make you wonder why we use the word 'gene' at all. Why not use a collective noun like 'gene complex'? The answer is that for many purposes that is indeed quite a good idea. But if we look at things in another way, it does make sense too to think of the gene complex as being divided up into discrete replicators or genes. This arises because of the phenomenon of sex.

Sexual reproduction has the effect of mixing and shuffling genes. This means that any one individual body is just a temporary vehicle for a short-lived combination of genes. The combination of genes which is any one individual may be short-lived, but the genes themselves are potentially very long-lived. Their paths constantly cross and recross down the generations. One gene may be regarded as a unit which survives through a large number of successive individual bodies. This is the central argument which will be developed in this chapter. It is an argument which some of my most respected colleagues obstinately refuse to agree with, so you must forgive me if I seem to labour it! First, I must briefly explain the facts of sex.

I said that the plans for building a human body are spelt out in 46 volumes. In fact this was an over-simplification. The truth is rather bizarre. The 46 chromosomes consist of 23 pairs of chromosomes. We might say that, filed away in the nucleus of every cell, are two alternative sets of 23 volumes of plans. Call them Volume 1a and Volume 1b, Volume 2a and 2b etc., down to Volume 23 a and Volume 23b. Of course the identifying numbers I use for volumes and, later, pages, are purely arbitrary.

We receive each chromosome intact from one of our two parents, in whose testis or ovary it was assembled. Volumes 1a, 2a, 3a, . . . came, say, from the father. Volumes 1b, 2b, 3b, . . . came from the mother. It is very difficult in practice, but in theory you could look with a microscope at the 46 chromosomes in any one of your cells, and pick out the 23 that came from your father and the 23 that came from your mother. The paired chromosomes do not spend all their lives physically in contact with each other, or even near each other. In what sense are they 'paired'? In the sense that each volume coming originally from the father can be regarded, page for page, as a direct alternative to one particular volume coming originally from the mother. For instance, Page 6 of Volume 13a and Page 6 of Volume 13b might both be 'about' eye colour; perhaps one says 'blue' while the other says 'brown.'

Sometimes the two alternative pages are identical, but in other cases, as in our example of eye colour, they differ. If they make contradictory 'recommendations,' what does the body do? The answer varies. Sometimes one reading prevails over the other. In the eye colour example just given, the person would actually have brown eyes: the instructions for making blue eyes would be ignored in the building of the body, though this does not stop them being passed on to future generations. A gene which is ignored in this way is called "recessive." The opposite of a recessive gene is a "dominant" gene. The gene for brown eyes is dominant to the gene for blue eyes. A person has blue eyes only if both copies of the relevant page are unanimous in recommending blue eyes. More usually when two alternative genes are not identical, the result is some kind of compromise—the body is built to an intermediate design or something completely different. When two genes, like the brown eye and blue eye gene, are rivals for the same slot on a chromosome, they are called "alleles" of each other.

Our genes are doled out to us at conception, and there is nothing we can do about this. Nevertheless, there is a sense in which, in the long term, the genes of the population in general can be regarded as a "gene pool." This phrase is in fact a technical term used by geneticists. The gene pool is a worthwhile abstraction because sex mixes genes up, albeit in a carefully organized way. In particular, something like the detaching and interchanging of pages and wads of pages from loose-leaf binders really does go on.

I have described the normal division of a cell into two new cells, each one receiving a complete copy of all 46 chromosomes. This normal cell division is called "mitosis." But there is another kind of cell division called "meiosis." This occurs only in the production of the sex cells; the sperms or eggs. Sperms and eggs are unique among our cells in that, instead of containing 46 chromosomes, they contain only 23. This is, of course, exactly half of 46—convenient when they fuse in sexual fertilization to make a new individual! Meiosis is a special kind of cell division, taking place only in testicles and ovaries, in which a cell with the full double set of 46 chromosomes divides to form sex cells with the single set of 23 (all the time using human numbers for illustration).

Which 23 are put into any given sperm cell? It is clearly important that a sperm should not get just any old 23 chromosomes. It mustn't end up with two copies of Volume 13 and none of Volume 17. It would theoretically be possible for an individual to endow one of his sperms with chromosomes which came, say, entirely from his mother. In this unlikely event, a child conceived by the sperm would inherit half her genes from her paternal grandmother, and none from her paternal grandfather. But in fact this kind of gross, whole-chromosome distribution does not happen. The truth is rather more complex.

Remember that the volumes (chromosomes) are to be thought of as loose-leaf binders. What happens is that, during the manufacture of the sperm, single pages, or rather multi-page chunks, are detached and swapped with the corresponding chunks from the alternative volume. Therefore every sperm cell made by an individual is unique, even though all his sperms assembled their 23 chromosomes from bits of the same set of 46 chromosomes. Eggs are made in a similar way in ovaries, and they too are all unique.

The process of swapping bits of chromosome is called "crossing over." It is very important for the whole argument of this book. It means that if you got out your microscope and looked at the chromosomes in one of your own sperms (or eggs if you are female) it would be a waste of time trying to identify chromosomes which originally came from your father and chromosomes which originally came from your

My Grandparents, My Parents, and I (Family Tree) (Frida Kahlo, 1936)

mother. (This is in marked contrast to the case of ordinary body cells.) Any one chromosome in a sperm would be a patchwork, a mosaic of maternal genes and paternal genes.

Immortal Coils

The average life expectancy of a genetic unit can conveniently be expressed in generations, which can in turn be translated into years. If we take a whole chromosome as our presumptive genetic unit, its life story lasts for only one generation. Suppose it is your chromosome number 8a, inherited from your father. It was created inside one of your father's testicles, shortly before you were conceived. It had never existed before in the whole history of the world. It was created by the meiotic shuffling process, forged by the coming together of pieces of chromosome from your paternal grandmother and your paternal grandfather. It was placed inside one particular sperm, and it was unique. The sperm was one of several millions, a vast armada of tiny vessels, and together they sailed into your mother. This particular sperm was the only one of the flotilla which found harbour in one of your mother's eggs—that is why you exist.

The genetic unit we are considering, chromosome number 8a, set about replicating itself along with all the rest of your genetic material. Now it exists, in duplicate form, all over your body. But when you in your turn come to have children, this chromosome will be destroyed when you manufacture eggs (or sperms). Bits of it will be exchanged with your maternal chromosome number 8b. In any one sex cell, a new chromosome number 8 will be created, perhaps 'better' than the old one, perhaps 'worse,' but, barring a rather improbable coincidence, definitely different, definitely unique. The life-span of a chromosome is one generation.

Now comes the important point. The shorter a genetic unit is, the longer—in generations—it is likely to live. In particular, the less likely it is to be split by any one crossing-over. This means that the unit can expect to survive for a large number of generations in the individual's descendants.

I am using the word gene to mean a genetic unit which is small enough to last for a large number of generations and to be distributed around in the form of many copies. This is not a rigid all-or-nothing definition, but a kind of fading-out definition, like the definition of "big" or "old." What I have done is to define a gene as a unit which, to a high degree, approaches the ideal of indivisible particulateness. A gene is not indivisible, but it is seldom divided. It is either definitely present or definitely absent in the body of any given individual. A gene travels intact from grandparent to grandchild, passing straight through the intermediate generation without being merged with other genes.

Another aspect of the particulateness of the gene is that it does not grow senile; it is no more likely to die when it is a million years old than when it is only a hundred. It leaps from body to body down the generations, manipulating body after body in its own way and for its own ends, abandoning a succession of mortal bodies before they sink in senility and death.

The genes are the immortals, or rather, they are defined as genetic entities which come close to deserving the title. We, the individual survival machines in the world, can expect to live a few more decades. But the genes in the world have an expectation of life which must be measured not in decades but in thousands and millions of years.

An individual body seems discrete enough while it lasts, but alas, how long is that? Each individual is unique. You cannot get evolution by selecting between entities when there is only one copy of each entity! Sexual reproduction is not replication. Just as a population is contaminated by other populations, so an individual's posterity is contaminated by that of his sexual partner. Your children are only half you, your grandchildren only a quarter you. In a few generations the most you can hope for is a large number of descendants, each of whom bears only a tiny portion of you—a few genes—even if a few do bear your surname as well.

Individuals are not stable things, they are fleeting. Chromosomes too are shuffled into oblivion, like hands of cards soon after they are dealt. But the cards themselves survive the shuffling. The cards are the genes. The genes are not destroyed by crossing-over, they merely change partners and march on. They are the replicators and we are their survival machines. When we have served our purpose we are cast aside. But genes are denizens of geological time: genes last forever.

Genes, like diamonds, are forever, but not quite in the same way as diamonds. It is an individual diamond crystal which lasts, as an unaltered pattern of atoms. DNA molecules don't have that kind of permanence. The life of any one physical DNA molecule is quite short—perhaps a matter of months, certainly not more than one lifetime. But a DNA molecule could theoretically live on in the form of copies of itself for a hundred million years. Moreover, just like the ancient replicators, copies of a particular gene may be distributed all over the world. The difference is that the modern versions are all neatly packaged inside the bodies of survival machines.

It is its potential immortality that makes a gene a good candidate as the basic unit of natural selection. But now the time has come to stress the word "potential." A gene can live for a million years, but many new genes do not even make it past their first generation. The few new ones who succeed do so partly because they are lucky, but mainly because they have what it takes, and that means they are good at making survival machines. They have an effect on embryonic development of

C. G. E. Dawkins b. 1882

C. J. Dawkins b. 1915

"Your children are only half you, your grandchildren only a quarter you. In a few generations the most you can hope for is a large number of descendants, each of whom bears only a tiny portion of you—a few genes—even if a few do bear your surname as well."—Richard Dawkins

Richard Dawkins b. 1941

each successive body in which they find themselves, such that that body is a little bit more likely to live and reproduce than it would have been under the influence of the rival gene or allele. For example, a "good" gene might ensure its survival by tending to endow the successive bodies in which it finds itself with long legs, which help those bodies to escape predators. This is a particular example, not a universal one. Long legs, after all, are not always an asset. To a mole they would be a handicap.

Rather than bog ourselves down in details, can we think of any universal qualities which we would expect to find in all good (i.e. long-lived) genes? Conversely, what are the properties which instantly mark a gene out as a 'bad,' short-lived one? There might be several such universal properties, but there is one which is particularly relevant to this book: at the gene level, altruism must be bad and selfishness good. This follows inexorably from our definitions of altruism and selfishness. Genes are competing directly with their alleles for survival, since their alleles in the gene pool are rivals for their slot on the chromosomes of future generations. Any gene which behaves in such a way as to increase its own survival chances in the gene pool at the expense of its alleles will, by definition, tautologously, tend to survive. The gene is the basic unit of selfishness.

15 *Evolution of Consciousness*

For me, the most powerful passages in Richard Dawkins' book The Selfish Gene *explored the evolution of consciouness. Thus intrigued, I looked for more books and articles on the subject. I discovered that while Dawkins' selfish gene theory is as nearly original as any new theory in science can ever be, his ruminations on consciousness draw from a wide and deeply explored literature. Nevertheless, his work in this second area has made an impact in at least one way: he coined the word "meme," which is so simple, so delightful, and (as you will see) so relevant that it is now entrenched in the jargon of science.*

One aspect of the evolution of consciousness that Dawkins investigates is "cultural evolution." Ashley Montagu (whose critique of sociobiology appeared earlier) and Theodosius Dobzhansky together legitimized the notion of cultural evolution in the years immediately following World War II. The timing is not coincidental. Horrified by the Nazi racial prejudice and quest for genetic purity that gave rise to the Holocaust, socially motivated scientists searched for a way in which humankind could overcome the dictates of the genes. Their answer: cultural evolution.

Programming Behavior

Richard Dawkins

Both animals and plants evolved into many-celled bodies, complete copies of all the genes being distributed to every cell. Some people use the metaphor of colony, describing the body as a colony of cells. I prefer to think of the body as a colony of genes, and of the cell as a convenient working unit for the chemical industries of the genes.

Colonies of genes they may be but, in their behaviour, bodies have undeniably acquired an individuality of their own. An animal moves as a coordinated whole, as a unit. Subjectively I feel like a unit, not a colony. This is to be expected. Selection has favoured genes which cooperate with others. In the fierce struggle to eat other survival machines, and to avoid being eaten, there must have been a premium on central coordination rather than anarchy with the communal body. Nowadays the intricate mutual co-evolution of genes has proceeded to such an extent that the communal nature of an individual survival machine is virtually unrecognizable. Indeed many biologists do not recognize it, and will disagree with me.

One of the most striking properties of survival machine behavior is its apparent purposiveness. By this I do not just mean that it seems to be well calculated to help the animal's genes to survive, although of course it is. I am talking about a closer analogy to human purposeful behaviour. When we watch an animal 'searching' for food, or for a mate, or for a lost child, we can hardly help imputing to it some of the subjective feelings we ourselves experience when we search. Each one of us knows, from the evidence of his own introspection, that, at least in one modern survival machine, this purposiveness has evolved the property we call "consciousness." I am not philosopher enough to discuss what this means, but fortunately it does not matter for our present purposes because it is easy to talk about machines which behave as if motivated by a purpose, and to leave open the question whether they are actually conscious.

The genes control the behaviour of their survival machines, not directly with their fingers on puppet strings, but indirectly like the computer programmer. All they can do is to set it up beforehand; then the survival machine is on its own, and the genes can only sit passively inside. Why are they so passive? Why don't they grab the reins and take charge from moment to moment? The answer is that they cannot because of time-lag problems.

Genes work by controlling protein synthesis. This is a powerful way of manipulating the world, but it is slow. It takes months of patiently pulling protein strings to build an embryo. The whole point about behaviour, on the other hand, is that it is fast. It works on a time-scale not of months but of seconds and fractions of seconds. Something happens in the world, an owl flashes overhead, a rustle in the long grass betrays prey, and in milliseconds nervous systems crackle into action, muscles leap, and someone's life is saved—or lost. Genes don't have reaction times like that.

Learning, Simulation, and Self-Awareness

The genes can only do their best in advance by building a fast executive computer for themselves, and programming it in advance with rules and 'advice' to cope with as many eventualities as they can 'anticipate.' But life, like the game of chess, offers too many different possible eventualities for all of them to be anticipated. Like the chess programmer, the genes have to 'instruct' their survival machines not in specifics, but in the general strategies and tricks of the living trade.

One way for genes to solve the problem of making predictions in rather unpredictable environments is to build in a capacity for learning. Here the program may take the form of the following instructions to the survival machine: "Here is a list of things defined as rewarding: sweet taste in the mouth, orgasm, mild temperature, smiling child. And here is a list of nasty things: various sorts of pain, nausea, empty stomach, screaming child. If you should happen to do something which is followed by one of the nasty things, don't do it again, but on the other hand repeat anything which is followed by one of the nice things." The advantage of this sort of programming is that it greatly cuts down the number of detailed rules which have to be built into the original program; and it is also capable of coping with changes in the environment which could not have been predicted in detail.

One of the most interesting methods of predicting the future is simulation. You imagine what would happen if you did each of the alternatives open to you. You set up a model in your head, not of everything in the world, but of the restricted set of entities which you think may be relevant. Survival machines which can simulate the future are one jump ahead of survival machines who can only learn on the basis of overt trial and error.

The evolution of the capacity to simulate seems to have culminated in subjective consciousness. Why this should have happened is, to me, the most profound mystery facing modern biology. There is no reason to suppose that electronic computers are conscious when they simulate, although we have to admit that in the future they may become so. Perhaps consciousness arises when the brain's simulation of the world becomes so complete that it must include a model of itself.

Whatever the philosophical problems raised by consciousness, for the purpose of this story it can be thought of as the culmination of an evolutionary trend towards the emancipation of survival machines as executive decision-takers from their ultimate masters, the genes. Not only are brains in charge of the day-to-day running of survival-machine affairs, they have also acquired the ability to predict the future and act accordingly. They even have the power to rebel against the dictates of the genes, for instance in refusing to have as many children as they are able to. But in this respect man is a very special case.

Cultural Evolution

Most of what is unusual about man can be summed up in one word: culture. I use that word not in its snobbish sense, but as a scientist uses it. Cultural transmission is analogous to genetic transmission in that, although basically conservative, it can give rise to a form of evolution. Geoffrey Chaucer could not hold a conversation with a modern Englishman, even though they are linked to each other by an unbroken chain of some twenty generations of Englishmen, each of whom could speak to his immediate neighbours in the chain as a son speaks to his father. Language seems to 'evolve' by non-genetic means, and at a rate which is orders of magnitude faster than genetic evolution.

Fashions in dress and diet, ceremonies and customs, art and architecture, engineering and technology, all evolve in historical time in a way that looks like highly speeded up genetic evolution, but has really nothing to do with genetic evolution. As in genetic evolution though, the change may be progressive. There is a sense in which modern science is actually better than ancient science. Not only does our understanding of the universe change as the centuries go by: it improves. Admittedly the current burst of improvement dates back only to the Renaissance, which was preceded by a dismal period of stagnation, in which European scientific culture was frozen at the level achieved by the Greeks. But evolution too may proceed as a series of brief spurts between stable plateaux.

As an enthusiastic Darwinian, I have been dissatisfied with explanations which my fellow-enthusiasts have offered for human behaviour. They have tried to look for "biological advantages" in various attributes of human civilization. For instance, tribal religion has been seen as a mechanism for solidifying group identity, valuable for a pack-hunting species whose individuals rely on cooperation to catch large and fast prey. "Kin selection" and selection in favour of reciprocal altruism may have acted on human genes to produce many of our basic psychological attributes and tendencies. These ideas are plausible as far as they go, but I find that they do not begin to square up to the formidable challenge of explaining culture, cultural evolution, and the immense differences between human cultures around the world, from the utter selfishness of the Ik of Uganda, as described by Colin Turnbull, to the gentle altruism of Margaret Mead's Arapesh.

I think we have got to start again and go right back to first principles. The argument I shall advance, surprising as it may seem coming from the author of the earlier chapters, is that, for an understanding of the evolution of modern man, we must begin by throwing out the gene as the sole basis of our ideas on evolution. I am an enthusiastic Darwinian, but I think Darwinism is too big a theory to be confined to the narrow context of the gene.

What after all is so special about genes? The answer is that they are replicators. The laws of physics are supposed to be true all over the accessible universe. Are there any principles of biology which are likely to have similar universal validity? When astronauts voyage to distant planets and look for life, they can expect to find creatures too strange and unearthly for us to imagine. But is there anything which must be true of all life, wherever it is found, and whatever the basis of its chemistry?

Obviously I do not know but, if I had to bet, I would put my money on one fundamental principle. This is the law that all life evolves by the differential survival of replicating entities. The gene, the DNA molecule, happens to be the replicating entity which prevails on our own planet. There may be others. If there are, provided certain other conditions are met, they will almost inevitably tend to become the basis for an evolutionary process.

But do we have to go to distant worlds to find other kinds of replicator and other, consequent, kinds of evolution? I think that a new kind of replicator has recently emerged on this very planet. It is staring us in the face. It is still in its infancy, still drifting clumsily about in its primeval soup, but already it is achieving evolutionary change at a rate which leaves the old gene panting far behind.

Memes: The New Replicators

The new soup is the soup of human culture. We need a name for the new replicator, a noun which conveys the idea of a unit of cultural transmission, or a unit of imitation. "Mimeme" comes from a suitable Greek root, but I want a monosyllable that sounds a bit like "gene." I hope my classicist friends will forgive me if I abbreviate mimeme to "meme." If it is any consolation, it could alternatively be thought of as being related to "memory," or to the French word "même" *(which means "same").* It should be pronounced to rhyme with "cream."

Examples of memes are tunes, ideas, catch-phrases, clothes, fashions, ways of making pots or of building arches. Just as genes propagate themselves in the gene pool by leaping from body to body via sperms or eggs, so memes propagate themselves in the meme pool by leaping from brain to brain via a process which, in the broad sense, can be called imitation. If a scientist hears or reads about a good idea, he passes it on to his colleagues and students. He mentions it in his articles and his lectures. If the idea catches on, it can be said to propagate itself, spreading from brain to brain. As my colleague N. K. Humphrey neatly summed up an earlier draft of this chapter: "Memes should be regarded as living structures, not just metaphorically but technically. When you plant a fertile meme in my mind you literally parasitize my brain, turning it into a vehicle for the meme's propagation in just the way that a virus may parasitize the genetic mechanism of a host cell."

For more than three thousand million years, DNA has been the only replicator worth talking about in the world. But it does not necessarily hold these monopoly rights for all time. Whenever conditions arise in which a new kind of replicator can make copies of itself, the new replicators will tend to take over, and start a new kind of evolution of their own. Once this new evolution begins, it will in no necessary sense be subservient to the old. The old gene-selected evolution, by making brains, provided the "soup" in which the first memes arose. Once self-copying memes had arisen, their own, much faster, kind of evolution took off. We biologists have assimilated the idea of genetic evolution so deeply that we tend to forget that it is only one of many possible kinds of evolution.

Some memes, like some genes, achieve brilliant short-term success in spreading rapidly, but do not last long in the meme pool. Popular songs and stiletto heels are examples. Others, such as the Jewish religious laws, may continue to propagate themselves for thousands of years, usually because of the great potential permanence of written records.

At first sight it looks as if memes are not high-fidelity replicators at all. Every time a scientist hears an idea and passes it on to somebody else, he is likely to change it somewhat. I have made no secret of my debt in this book to the ideas of Robert L. Trivers. Yet I have not repeated them in his own words. I have twisted them round for my own purposes, changing the emphasis, blending them with ideas of my own and of other people. The memes are being passed on to you in altered form. This looks quite unlike the particulate, all-or-none quality of gene transmission. It looks as though meme transmission is subject to continuous mutation, and also to blending.

When we say that all biologists nowadays believe in Darwin's theory, we do not mean that every biologist has, graven in his brain, an identical copy of the exact words of Charles Darwin himself. Each individual has his own way of interpreting Darwin's ideas. He probably learned them not from Darwin's own writings, but from more recent authors. Much of what Darwin said is, in detail, wrong. Darwin if he read this book would scarcely recognize his own original theory in it, though I hope he would like the way I put it. Yet, in spite of all this, there is something, some essence of Darwinism, which is present in the head of every individual who understands the theory. The meme of Darwin's theory is therefore that essential basis of the idea which is held in common by all brains who understand the theory.

Revolt of the Survival Machines

I have been a bit negative about memes, but they have their cheerful side as well. When we die there are two things we can leave behind us: genes and memes. We were built as gene machines, created to pass on our genes. But that aspect of us will be forgotten in three generations. Your child, even your grandchild, may bear a resemblance to you, perhaps in facial features, in a talent for music, in the colour of her hair. But as each generation passes, the contribution of your genes is halved. It does not take long to reach negligible proportions. Our genes may be immortal but the collection of genes which is any one of us is bound to crumble away. Elizabeth II is a direct descendent of William the Conqueror. Yet it is quite probable that she bears not a single one of the old king's genes. We should not seek immortality in reproduction.

But if you contribute to the world's culture, if you have a good idea, compose a tune, invent a sparking plug, write a poem, it may live on, intact, long after your genes have dissolved in the common pool. Socrates may or may not have a gene or two alive in the world today, but who cares? The meme-complexes of Socrates, Leonardo, Copernicus, and Marconi are still going strong.

"The heavy infantry of rebellious insects" (Grandville, early 19th century)

I now close the topic of the new replicators, and end the book on a note of qualified hope. Even if we look on the dark side and assume that individual man is fundamentally selfish, our conscious foresight—our capacity to simulate the future in imagination—could save us from the worst selfish excesses of the blind replicators. We have at least the mental equipment to foster our long-term selfish interests rather than merely our short-term selfish interests. We have the power to defy the selfish genes of our birth and, if necessary, the selfish memes of our indoctrination. We can even discuss ways of deliberately cultivating and nurturing pure, disinterested altruism—something that has no place in nature, something that has never existed before in the whole history of the world. We are built as gene machines and cultured as meme machines, but we have the power to turn against our creators. We, alone on earth, can rebel against the tyranny of the selfish replicators.

Dawkins weaves a compelling argument, but you may have noticed that his case is based more on logic than on tangible evidence. Certainly, the pioneering work of William Hamilton revealing kin selection in colonial insects is strong support, as is Robert Trivers' research on reciprocal altruism, and later developments in game theory provide powerful tools. But is there anything else that supports this radical theory?

The Extended Phenotype

Six years after The Selfish Gene *was published, Dawkins produced* The Extended Phenotype *(W. H. Freeman, 1982). There he cites more evidence, notably the fact that a gene ("genotype") contained in the body ("phenotype") of one organism can affect the body or behavior of a quite different species. The 1989 edition of* The Selfish Gene *includes a new chapter that summarizes the insights and examples drawn from this more technical book. Dawkins cites several examples of the power of extended phenotypes. For me, the strangest is this:*

Richard Dawkins

The case of the fluke ("brainworm") *Dicrocoelium dendriticum* is often quoted as another elegant example of a parasite manipulating an intermediate host to increase its chances of ending up in its definitive host. The definitive host is an ungulate such as a sheep, and the intermediate hosts are first a snail and then an ant. The normal life cycle calls for the ant to be accidentally eaten by the sheep. By burrowing into the suboesophageal ganglion, the aptly named "brainworm" changes the ant's behavior. Whereas an uninfected ant would normally retreat into its nest when it became cold, infected ants climb to the top of grass stems, clamp their jaws in the plant and remain immobile as if asleep. Here they are vulnerable to being eaten by the worm's definitive host. The infected ant, like a normal ant, does retreat down the grass stem to avoid death from the midday heat—which would be bad for the parasite—but it returns to its aerial resting position in the cool of the afternoon.

As a child, I was familiar with one very common example of parasitic manipulation of another body (though, of course, I didn't perceive it as such). In the weedlot out back I searched for swellings on plant stems and leaves, "galls," breaking these open to reveal insect larvae housed within. The insect genes somehow induce plant tissues to swell in the spot where the egg has been laid. The plant thereby supplies the growing grub not only with food but also with a safe and comfortable home. In his seventeenth-century treatise The Anatomy of Plants, *Marcello Malpighi revealed the cause and nature of plant galls thus:*

Marcello Malpighi

Nature has so arranged things that not only the higher animals should provide food for each other . . . but that also insects inherit in the plants a fertile breeding place, and nature has endowed them with such ingenuity that they force the plants to provide the uterus and, so to speak, the nourishing breasts for the eggs they lay on them. This service provided by the plants results in their own disfigurement in that they often develop a disease in the form of a swelling to which we like to give the name "galls."

Overall, with the extended phenotype idea Dawkins has enriched his earlier selfish-gene theory. For if the phenotypic effects of a gene can extend beyond the boundaries of its own survival machine, perhaps genes are entities that are more real than are organisms. Perhaps the selfish gene theory is true.

Viruses

Viruses too can be viewed as support for Dawkins' theory. They are the closest thing to simple, naked replicators that can be found on the planet today. Indeed, there is some dispute as to whether viruses should even be deemed alive.

The simplest bacterium is extraordinarily more complex than a virus, which is no more than a strand of nucleic acid (DNA or RNA) covered by a protective coat. A virus is the ultimate parasite because outside of a host cell it can do absolutely nothing. But once inside, it hijacks the cell's own metabolic machinery for the purpose of producing more strands of the viral genes. Production of viral genes goes on unimpeded until the volume of product is so great that the cell walls burst and thus release thousands or millions of copies of the virus into the surrounding tissues or bloodstream to begin the process anew.

Viruses, therefore, do none of the usual things we associate with living organisms. They don't eat; they don't breathe; they don't grow. All they do is replicate. There is no doubt that from a virus's point of view the entire meaning of life is replication.

Natural Selection at the Gene Level

Perhaps the most compelling evidence in support of the selfish gene theory appeared in the 17 April 1980 issue of Nature *284:601–607. The journal published two papers by prominent biologists: one by W. Ford Doolittle and Carmen Sapienza and the other by Leslie Orgel and Francis Crick. Doolittle is a molecular biologist at Dalhousie University in the Canadian province of Nova Scotia. (You may recall Doolittle's*

Seventeenthth-century botanist Marcello Malpighi discovered that plant galls were the work not of witchcraft but of insect parasites. In his artistic rendering of a Mediterranean oak gall, Malpighi reveals the larva that grew from an egg inserted into the plant by the long ovipositor of a wasp.

critique of the Gaia hypothesis in chapter 2.) Sapienza was Doolittle's graduate student in 1980 and is now head of the Laboratory of Developmental Genetics at the Ludwig Institute for Cancer Research in Montreal. Orgel conducted pioneering research in the origins of life, specifically the prebiotic synthesis of nucleic acids. Crick, of course, is famous for his codiscovery with James Watson of the helical structure of DNA.

Working independently, these two teams reported discovering "selfish DNA" or "junk DNA." They found that chromosomes are studded with multiple copies of genes and that, moreover, these spare copies don't seem to do anything useful for the organism. This excess genetic material is not transcribed into proteins. It simply exists, and it is replicated each time the chromosome as a whole replicates.

Following are excerpts (minus the citations) from these two groundbreaking scientific papers (reprinted by permission of the authors and Nature, *copyright 1980 by Macmillan Magazines Ltd.). The authors wrote for a wide audience, not just molecular biologists and evolution theorists. The writing is therefore lively and relatively free of jargon. The journal too was aware of the importance of this work. Had these two papers not appeared in precisely the same weekly issue of* Nature, *only one of the teams would have captured most of the glory. The premium placed on being first to publish has often been criticized for inducing secretiveness, hostility, and paranoia among scientists. But it also induces scientists to get their ideas out, so that other scientists can work on them too. In this case, both teams acknowledge advice and support from the other.*

W. Ford Doolittle, Carmen Sapienza

The assertion that organisms are simply DNA's way of producing more DNA has been made so often that it is hard to remember who made it first. Certainly, Dawkins has provided the most forceful and uncompromising recent statement of this position, as well as of the position that it is the gene, and not the individual or the population, upon which natural selection acts. Although we may thus view genes and DNA as essentially selfish, most of us are, nevertheless, wedded to what we will call here the "phenotype paradigm"—the notion that the major and perhaps only way in which a gene can ensure its own perpetuation is by ensuring the perpetuation of the organism it inhabits.

The phenotype paradigm underlies attempts to explain genome structure. There is a hierarchy of types of explanations we use in efforts to rationalize, in neo-darwinian terms, DNA sequences which do not code for protein. Untranslated messenger RNA sequences which precede, follow or interrupt protein-coding sequences are often assigned a phenotypic role in regulating messenger RNA maturation, transport or translation. Such interpretations of DNA structure are very often demonstrably correct; molecular biology would not otherwise be so fruitful. However, the phenotype paradigm is almost tautological; natural selection operates on DNA through organismal phenotype, so DNA structure must be of immediate or long-term (evolutionary) phenotypic benefit, even when we cannot show how. As Gould and Lewontin note, "the rejection of one adaptive story usually leads to its replacement by another, rather than to a suspicion that a different kind of explanation might be required. Since the range of adaptive stories is as wide as our minds are fertile, new stories can always be postulated."

What we propose here is that there are classes of DNA for which a "different kind of explanation" may well be required. Natural selection does not operate on DNA only through organismal phenotype. Cells themselves are environments in which DNA sequences can replicate, mutate and so evolve. Although DNA sequences which contribute to organismal phenotypic fitness or evolutionary adaptability indirectly increase their own chances of preservation, and may be maintained by classical phenotypic selection, the only selection pressure which DNAs experience directly is the pressure to survive within cells. If there are ways in which mutation can increase the probability of survival within cells without effect on organismal phenotype, then sequences whose only "function" is self-preservation will inevitably arise and be maintained by what we call "non-phenotypic selection." Furthermore, if it can be shown that a given gene (region of DNA) or class of genes (regions) has evolved a strategy which increases its probability of survival within cells, then no additional (phenotypic) explanation for its origin or continued existence is required.

Selfish DNA: The Ultimate Parasite

Leslie Orgel,
Francis Crick

The object of this short review is to make widely known the idea of selfish DNA. A piece of selfish DNA, in its purest form, has two distinct properties:

- It arises when a DNA sequence spreads by forming additional copies of itself within the genome.
- It makes no specific contribution to the phenotype.

This idea is not new. We have not attempted to trace it back to its roots. It is sketched briefly but clearly by Dawkins in his book *The Selfish Gene*. We shall use the term "selfish DNA" in a wider sense, so that it can refer not only to obvious repetitive DNA but also to certain other DNA sequences which appear to have little or no function, such as much of the DNA in the introns of genes and parts of the DNA sequences between genes. Doolittle and Sapienza have independently arrived at similar ideas.

The large amounts of DNA in the cells of most higher organisms and, in particular, the exceptionally large amounts in certain animal and plant species has been an unsolved puzzle for a considerable period. As is well known, this DNA consists in part of "simple" sequences [that] are usually not transcribed. There is also evidence to suggest that the majority of DNA sequences in most higher organisms do not code for protein since they do not occur at all in messenger RNA. Nor is it very plausible that all this extra DNA is needed for gene control, although some portion of it certainly must be.

We also have to account for the vast amount of DNA found in certain species, such as lilies and salamanders, which may amount to as much as twenty times that found in the human genome. It seems totally implausible that the number of radically different genes needed in a salamander is twenty times that in a man. The conviction has been growing that much of this extra DNA is "junk," in other words, that it has little specificity and conveys little or no selective advantage to the organism.

The theory of natural selection, in its more general formulation, deals with the competition between replicating entities. It shows that, in such a competition, the more efficient replicators increase in number at the expense of their less efficient competitors. After a sufficient time, only the most efficient replicators survive. The idea of selfish DNA is firmly based on this general theory of natural selection, but it deals with selection in an unfamiliar context.

The familiar neo-darwinian theory of natural selection is concerned with the competition between organisms in a population. At the level of molecular genetics it provides an explanation for the spread of "useful" genes or DNA sequences within a population. Organisms that carry a

“My pet salamander has 20 times more DNA than I have.”—Connie Barlow

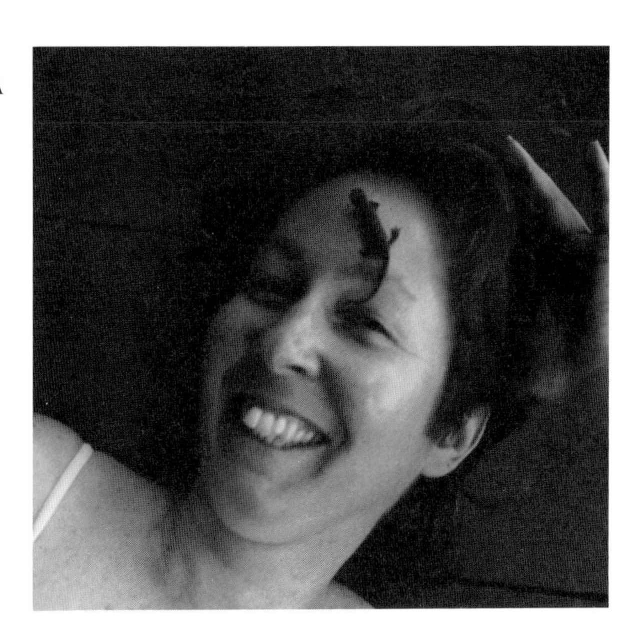

gene that contributes positively to fitness tend to increase their representation at the expense of organisms lacking that gene. In time, only those organisms that carry the useful gene survive. Natural selection also predicts the spread of a gene or other DNA sequence within a single genome, provided certain conditions are satisfied. If an organism carrying several copies of the sequence is fitter than an organism carrying a single copy, and if mechanisms exist for the multiplication of the relevant sequence, then natural selection must lead to the emergence of a population in which the sequence is represented several times in every genome.

The idea of selfish DNA is different. It is again concerned with the spread of a given DNA within the genome. However, in the case of selfish DNA, the sequence which spreads makes no contribution to the phenotype of the organism, except insofar as it is a slight burden to the cell that contains it. The spread of selfish DNA sequences within the genome can be compared to the spread of a not-too-harmful parasite within its host.

Some selfish DNA may acquire a useful function and confer a selective advantage on the organism. Using the analogy of parasitism, slightly harmful infestation may ultimately be transformed into a symbiosis. What we would stress is that not all selfish DNA is likely to become useful. Much of it may have no specific function at all. It would be folly in such cases to hunt obsessively for one. To continue our analogy, it is difficult to accept the idea that all human parasites have been selected by human beings for their own advantage.

The theory of selfish DNA is not so vague that it cannot be tested. We can think of three general ways to do this. In the first place, it is important to know where DNA sequences occur which appear to have little obvious function, whether they are associated with flanking or other sequences of any special sort and how homologous sequences differ in different organisms and in different species, either in sequence or in position on the chromosome. Second, if the increase of selfish DNA and its movement around the chromosome are not rare events in evolution, it may be feasible to study, in laboratory experiments, the actual molecular mechanisms involved in these processes. Third, one would hope that a careful study of all the nonspecific effects of extra DNA would give us a better idea of how it affected different aspects of cellular behaviour. In particular, it is important to discover whether the addition of nonspecific DNA does, in fact, slow down cells metabolically and for what reasons. Such information, together with a careful study of the physiology and life style of related organisms with dissimilar amounts of DNA, should eventually make it possible to explain these differences in a convincing way.

While proper care should be exercised both in labelling as selfish DNA every piece of DNA whose function is not immediately apparent and in invoking plausible but unproven hypotheses concerning the details of natural selection, the idea seems a useful one to bear in mind when exploring the complexities of the genomes of higher organisms. It could well make sense of many of the puzzles and paradoxes which have arisen over the last ten or fifteen years. The main facts are, at first sight, so odd that only a somewhat unconventional idea is likely to explain them.

Judging the Selfish Gene Theory

In comparing the gene's-eye view with biology's traditional focus on the organism, John Maynard Smith observed:

John Maynard Smith

It would be as foolish to argue about which of these views is correct as it would be to argue about whether algebra or geometry is the correct way to solve problems in science. It all depends on the problem you are trying to solve.

Thus John Maynard Smith evaluates the selfish gene theory from the standpoint of utility rather than truth, and he finds that it can indeed be useful. In that sense, both the selfish gene theory and the Gaia hypothesis are good science. But how should we nonscientists view Dawkins' thesis—we who may be more interested in truth than utility? A century ago the philosopher and psychologist William James made a statement that eerily anticipates Dawkins' worldview:

William James

The entire modern deification of survival per se, survival returning to itself, survival naked and abstract, with the denial of any substantive excellence in *what* survives, except the capacity for more survival still, is surely the strangest intellectual stopping-place ever proposed by one man to another.

If your head and your heart are split, if you find Richard Dawkins persuasive but William James nevertheless strikes a chord, you could find solace in Arthur Koestler's ideas. You can perhaps view Dawkins' theory as being the truth, but not the whole truth. Remember that it was Koestler who observed, "The members of a hierarchy, like the Roman god Janus, all have two faces looking in opposite directions: the face turned towards the subordinate levels is that of a self-contained whole; the faced turned upward towards the apex, that of a dependent part." Until recently, biologists have viewed genes as entirely dependent parts; Richard Dawkins now presses his colleagues to pay attention to genes as self-contained wholes.

While Stephen Jay Gould applauds the work of Doolittle and Sapienza, Orgel and Crick, he challenges some aspects of the broader selfish gene theory. One of his harshest criticisms concerns packaging rather than content:

Stephen Jay Gould

Selfish DNA is about the worst possible name for the phenomenon, for it records the very prejudice that the new structure of explanation should be combating: the exclusive focus on bodies as evolutionary agents. When we call repetitive DNA "selfish," we imply that it is acting for itself when it should be doing something else, namely, helping bodies in their evolutionary struggle. But are bodies the only kind of legitimate individuals in biology? Might there not be a hierarchy of individuals, with legitimate categories both above and below bodies: genes below, species above?

Along these lines David Hull, a philosopher of science, has suggested one fundamental change in outlook and vocabulary that would perhaps make the selfish gene theory less contentious scientifically and less offensive to human sensibilities. (Reprinted from Science as a Process *by David L. Hull, copyright 1988 by University of Chicago Press.)*

David Hull

Dawkins introduced the terms "replicator" and "vehicle" because of their generality and because of the common connotations of such terms as "gene" and "organism." But the terms he devised also have connotations. As far as I can see, the connotations of the term "replicator" are entirely appropriate, while those of "vehicle" are not. Vehicles are the sort of thing that agents ride around in. More than this, the agents are in control. They steer, and the vehicles follow dumbly. Although Dawkins explicitly assigns distinct evolutionary roles to both replicators and vehicles, the picture that Dawkins' terminology elicits is that of genes controlling helpless and hapless organisms. Thus, in an effort to reduce conceptual confusion, I suggest the following definitions:

• *replicator:* an entity that passes on its structure largely intact in successive replications.

• *interactor:* an entity that interacts as a cohesive whole with its environment in such a way that this interaction causes replication to be differential.

With the aid of these two technical terms, selection can be characterized succinctly as follows:

• *selection:* a process in which the differential extinction and proliferation of interactors cause the differential perpetuation of the relevant replicators.

Replicators and interactors are the entities that function in selection processes. Some general term is also needed for the entities that result from successive replications:

• *lineage:* an entity that persists indefinitely through time either in the same or an altered state as a result of replication.

As simple as the distinction between replicators and interactors is, it is sufficient to eliminate a recurrent disagreement about selection processes by showing that it is only apparent, not real. At least sometimes gene selectionists and organism selectionists are not disagreeing with each other but referring to two different aspects of the same process.

Overall, Richard Dawkins and his fellow sociobiologists have made a profound contribution to biology and to our worldview by showing us a gene's-eye impression of life. The gene is an entity itself—a holon, if you like—not just a part of the organism. While genes surely do 'serve' us (we wouldn't be here without them), we must recognize that they probably have their own agenda too, and it may not always be compatible with ours. In this way the reductionistic approach of Dawkins and colleagues complements the holistic approach of Lovelock and fellow adherents of a systems view of life—who encouraged us to look beyond the organism in the other direction, to the biosphere as a living entity itself.

VII *Reflections*

But the greatest thing by far is to be a master of metaphor.

—Aristotle (*Poetics*)

Lovelock offers a myth to express our wonder and gratitude, and Gaia is the muse of many who care deeply about this planet. Is it mean-spirited to question such a socially useful metaphor?

—W. Ford Doolittle, 1991

17 *The Debates Continue*

Six of the scientists and writers excerpted in this book submitted short essays expressly for this closing chapter.

Questioning a Metaphor

W. Ford Doolittle

Holists and reductionists don't often lie so peacefully between the same covers, although many of us of the latter ilk *do* accept that nature is hierarchical, and that parts make up wholes which show behaviors parts cannot. Atoms for instance cannot evolve by natural selection, whereas all replicating "individuals" (pieces of DNA, real genes, organisms, some kinds of populations, even "memes") that show heritable variation in fitness in principle can. The "selfish DNA" debate resulted, I believe, in agreement that chromosomes provide an environment in which DNAs can compete with each other through various replicative strategies, even to the detriment of organisms, and legitimized the notion of individuals and selection at the lowest level of the hierarchy. There are now very clear examples of this, but we still don't know how much of a genome (ours, for instance) is truly selfish, how much is useful, how much "junk." At the other end of the ladder, arguments about how high up the hierarchy we can truly find replicating entities (are species individuals?) still separate reductionists from holists, as do disagreements about whether there are principles other than natural selection which describe evolution at higher levels. At the risk of appearing surly, I'd like to devote these few words to explaining why I believe that there are no replicating entities at the highest level—the biosphere—and that there are no new general principles of evolutionary biology to be discovered there.

The biosphere, Lovelock's Gaia, is not a replicating individual, and has no coherent heredity. If her parts contribute to global homeostasis, this cannot be for the same reasons that the organs of an animal promote physiological homeostasis—not, that is, because there were ancient populations of proto-Gaias in which those with more self-control left progeny while others perished. If individual species do collaborate as one whole, it can only be because some interaction rule like Tit-for-Tat (see Axelrod and Hamilton, chapter 10) has been selected during their coevolution. It is very hard to imagine this happening when the consequences of defection by one species will not be perceived by others for generations or millenia. Reductionists prefer lower-level arguments. For instance, DMS may indeed control cloud formation (see Monastersky, chapter 2), but if DMS-deficient mutants grow better as individuals than wild-type cells, such mutants will take over plankton populations, and no higher-level authority can prevent this. The fact it hasn't happened means that DMS must be good for plankton cells as individuals, and given that, we need not invoke higher-level control.

The Gaia hypothesis may however not be about evolutionary biology and the levels of selection issue, but about ecology. I recall, as an undergraduate, wondering at the existence of carbon and nitrogen and sulfur cycles, and being grateful, since otherwise someday everything will pile up as the nonbiodegradable remains of the most luxuriant life forms. My invocation of the Anthropic Principle (see Doolittle, chapter 2) strikes some as jejune—as useful as marvelling that none of my own personal ancestors, not only back to Adam but to the first cell, died without issue.

Lovelock offers a myth to express our wonder and gratitude, and Gaia is the muse of many who care deeply about this planet. Is it mean-spirited to question such a socially useful metaphor? I like to use my own metaphor to express that moral dilemma:

It's as if we awoke in surprise to find ourselves at the controls of an over-booked jumbo jet hurtling through the night. We've never flown a plane before, but the warning lights have all gone red, and the passengers clamor for action. The passengers are divided, however, between those who think we are on automatic pilot (some even think that God is that automatic pilot) and those who fear we are already in a fatal nose-dive. Both camps agree we must learn what all those buttons are—the first so that we might not inadvertently assume manual control, the second so that we can do just that, before it is too late. Learn we must, even though we will not be able to decide between these two points of view until dawn breaks, and we can see whether or not the ground below us is littered with wreckage.

Biologists Can't Define Life

Lynn Margulis

More and more, like monasteries of the Middle Ages, today's universities and professional societies guard their knowledge. Collusively, the university biology curriculum, the textbook publishers, the National Science Foundation review committees, the graduate record examiners, and the various microbiological, evolutionary, and zoological societies map out domains of the known and knowable. They distinguish required from forbidden knowledge. Subtly punishing the trespassers with rejection and oblivion, they award the faithful liturgists by granting degrees and dispersing funds and fellowships. Biologists receive Guggenheim Fellowships for calculations of the evolutionary basis of altruism or quantification of parental investment in male children, while the tropical forests are destroyed at the rate of hundreds of acres per day and very little funding exists for study of live plants in their natural environments. It seems the English-speaking biology academy has lost sight of the biological priorities.

Furthermore, science practitioners today widely believe and teach—explicitly and by inference—that life is a mechanical system fully describable by physics and chemistry. Biology, in this reductionist view, is a subfield of chemistry and physics. Taken to its extreme, notably in the writings of Richard Dawkins, the machine-like image of "selfish replicator" becomes synonymous with life itself. I contrast this prevailing neo-Darwinian belief with a life-centered alternative worldview called "autopoiesis."

Autopoiesis was set forth as an alternative scientific paradigm in the 1970s by Chilean biologists Humberto Maturana and Francisco Varela. The name itself combines the Greek "auto" (self) and "poiesis" (to make); indeed, the term comes from the same root giving rise to poetry. It refers to the dynamic, self-producing, and self-maintaining activities of all living beings. The simplest, smallest known autopoietic entity is a single bacterial cell. The largest is probably Gaia. Cells and Gaia display general properties of autopoietic entities; as their surroundings change unpredictably, they maintain their structural integrity and internal organization, at the expense of solar energy, by remaking and interchanging their parts. Metabolism is the name given to this incessant activity.

For Vladimir Vernadsky, the early twentieth-century Russian scientist who pioneered the study of the biosphere, life is a special kind of mineral: animated water. Living entities are viewed as "problem solvers" by this century's most well-known philosopher of science, Karl Popper. Both Vernadsky and Popper thus recognize an aspect of living systems that is diminished by fitness-estimating, body-counting neo-Darwinists, and one that those espousing autopoiesis attempt to make explicit. In the autopoietic worldview, living beings do not merely adapt to a passive physico-chemical environment, as the neo-Darwinists assume. Rather, the activities of each organism lead to continuously changing environments. The oxygen we breathe, the humid atmosphere inside of which we live, and the mildly alkaline ocean waters in which the kelp and whales bathe are not determined by a physical universe run by mechanical laws; the surroundings are products of life interacting at the planet's surface.

Life is organized into individuals bounded by membranes and walls, including symbiotic mergers of such individuals. Plants and animals evolved from truces between predatory bacterial ancestors and their prey, while some cell organelles are relicts of undigested food. Fundamentally, life on Earth owes its long and continuing existence to these metabolic, physiological, behavioral, and evolutionary interactions. Gaia, 30 million types of organisms and the environmental consequences of their presence, *is* the Natural Selector.

Neo-Darwinists, some of whom painstakingly catalog the evolution of forms of animals and plants over the past 600 million years, virtually ignore the evolution of bacterial metabolic and genetic systems during the preceding 3000 million years; they never factor autopoiesis into their equations. Considering organisms as independent entities that evolve by accumulation of chance mutations, these scientists indeed must resist autopoiesis and a Gaian worldview. The current dilemma—the conflicting myths and thought-styles in professional biological science so vividly depicted in the contrasts between the first and second halves of this very book—is thus not likely to see resolution soon. But books like *From Gaia to Selfish Genes* can help in powerful ways; they can alert the reader, perhaps the next generation of biologists, to the fact that biologists have philosophies and that deep dilemmas exist.

This essay was adapted from "Big Trouble in Biology," in Doing Science: The Reality Club 2, *edited by John Brockman, Prentice Hall Press.*

The Way Forward

John Maynard Smith

This book leaves the reader with an unanswered question. How should biologists think about the relationship between wholes and parts—that is, between ecosystems and the species that compose them, between organisms and cells, or between cells and genes? Any attempt to explain biological complexity has to start with Darwin's theory of evolution by natural selection, because that is the only coherent biological theory we have. It asserts that, if there is a population of entities with the properties of multiplication, variation and heredity (like begets like), then those entities will evolve complex characteristics ensuring their survival and reproduction (that is, they will evolve adaptations).

Subsequent discussion has revealed that the relevant entities must have other properties if adaptive evolution is to occur: of these, the most important is that they should not themselves be composed of smaller entities, between which selection is acting. For example, if the cells of our bodies were competing to enter our gonads, and hence to transmit their genes to the next generation, we could not expect them to form organs, like hearts and kidneys, that function to ensure the survival of the whole organism. In this particular case, cooperation between cells has evolved because the body develops from a single cell, so that all the cells have (barring somatic mutation) identical genes; but the evolution of cooperation between parts is not actually so easily explained.

There are two classes of phenomena not readily explained by Darwin's theory. The first is the long-term survival of complex ecosystems: clearly, an ecosystem does not have the properties needed to evolve by natural selection. The second is the cooperation between entities (genes on a chromosome, organelles in a cell, cells in an organism, individuals in a society) that are today part of a larger whole, but were once independently evolving entities. Faced with these problems, people are liable to react in one of two ways, each, in my view, foolish. The first and commonest is to abandon Darwinism: that is like abandoning the atomic theory in chemistry because of nuclear fusion. The second is to deny the phenomenon, which is like asserting that nuclear fusion does not happen.

The way forward, I think, is to attempt a theory of the coevolution of many interacting species, and of the evolutionary origin of more complex organisms from simpler ones (prokaryotic cells from replicating molecules, eukaryotes from prokaryotes, animals and plants from protists, animal societies from solitary individuals) that is fully consistent with Darwinism. So far, we have had more success with the second of these projects than with the first, but there is much more to be done. I do not agree with the theories proposed by Lovelock, Margulis and Gould, but I am fascinated by the phenomena they discuss. What we need to do is to think about those phenomena with the hard-nosed reductionism of Axelrod, Hamilton and Dawkins.

In Defense of Writing

Dorion Sagan

In the nineteenth century Samuel Butler, a contemporary of Darwin, warned that scientists were the latest incarnation of the priest, the augur, the medicine man; although quite important to civilization, they must be watched awfully closely. The comments here by Lewontin, Rose, and Kamin on "science" echo such apprehension, and I was much pleased to read in them an image of my own concerns. Like Connie, merely a writer, I find that I am not in a privileged place to offer definitive pronouncements on truth—a role reserved in our society for the scientist—despite the paradox that "science" owes its authority precisely to the success of "the scientific method," which is supposed to hold in suspense all claims to truth while experimenting to prove various hypotheses. And yet this writerly disadvantage in the face of the scientific authority of others in this Reflections section suddenly turns to my advantage: as an onlooker or bystander I can be more objective—not about reality but about science, from which I have the required distance to form a perhaps clearer reflection, a less biased view.

In New Orleans recently the science writer Jon Weiner recently told me that the problem with science writing is that it is Abrahamic: always at a remove, journalistic, powerless like Abraham who in the Bible reports to the people not what God said but what Moses told him God said. I agree; the craving to empower writing attracts me to projects such as those of Lovelock, Koestler, and von Bertalanffy—individuals who, despite their profound differences, share the essentially Goethean goal of developing a "poetic science." But, as attested to by writers in the latter half of this book, science that allies itself with the poetic is considered by turns mystical, vitalist, new age. Labeled and dismissed as nonscience, poetic science only has a chance if its poetic elements go unrecognized, veiled. The need to keep the poetic roots of science hidden explains why Lovelock has gone from talking of "Gaia"—the Greek name for the Mother of the Titans—to "geophysiology"—a name in which the mythological nature of the Greek conception of the Earth (Gaea) is buried as the prefix (geo) of a scientific-sounding word.

But as a philosophically inclined writer I have found that it is possible to penetrate the magic circle scientists would draw around themselves to guard against "poetic" or "literary" incursions. All the scientists writing in this book are de facto writers which means that they must grapple with the unerasable ambiguity (and, indeed, inevitable evolution) of their words; if they choose not to, or pretend that there is an exact scientific truth to which their words imperfectly refer, I believe they have succumbed to a greater delusion than the "deconstructive" writer who has given up all hope of controlling meanings. The radical separation one must assume to distinguish between literal, concrete, univocal language on the one hand and writerly, equivocal language on the other is untenable. Indeed, the very labels used to validate non-metaphorical language—"clarity," "being concrete," "literal"—are themselves tropes, figures of speech. Clarity, for example, an image of transparency, refers to language in its ideality as a pure and undisturbing medium: not only is this untrue (especially in terms of opaque black marks on white paper) but the value of clarity is in direct opposition to the value of the concrete—which is, from a literal perspective, anything but "clear." So, too, "literal" in French is "propre," meaning "clean" as well as "own": there is much evidence to suggest that the very idea of a literal, nonfigurative, or "scientific" speech has to do with possession (the proper, property). What is one's own, what is true to itself, what is self-identical and in self-possession is not metaphorical.

Although these comments on language may seem confusing and off the point, they relate. In the transition from specialized discourse to pop-biology writing scientists reveal their humanity and, to the extent that they are good writers, they may undermine the mystique of science. This is good, since it allows us to move from an "Abrahamic" attitude in which we take on authority what scientists tell us, to a more enlightened one in which we begin to think things over for ourselves.

The Dilemmas

For his contribution to this final section, Edward O. Wilson chose to draw from the first chapter of his Pulitzer Prize-winning book, On

Five Kingdoms Hand (designed by Dorion Sagan)

Human Nature *(copyright 1978 by the President and Fellows of Harvard College and reprinted by permission of Harvard University Press).*

Edward O. Wilson

These are the central questions that the great philosopher David Hume said are of unspeakable importance: How does the mind work, and beyond that why does it work in such a way and not another, and from these two considerations together, what is man's ultimate nature? We keep returning to the subject with a sense of hesitancy and even dread. For if the brain is a machine of ten billion nerve cells and the mind can somehow be explained as the summed activity of a finite number of chemical and electrical reactions, boundaries limit the human prospect—we are biological and our souls cannot fly free. If humankind evolved by Darwinian natural selection, genetic chance and environmental necessity, not God, made the species. Deity can still be sought in the origin of the ultimate units of matter, in quarks and electron shells (Han Küng was right to ask atheists why there is something instead of nothing) but not in the origin of species. However much we embellish that stark conclusion with metaphor and imagery, it remains the philosophical legacy of the last century of scientific research.

No way appears around this admittedly unappealing proposition. It is the essential first hypothesis for any serious consideration of the human condition. Without it the humanities and social sciences are the limited descriptors of surface phenomena, like astronomy without physics, biology without chemistry, and mathematics without algebra. With it human nature can be laid open as an object of fully empirical research, biology can be put to the service of liberal education, and our self-conception can be enormously and truthfully enriched. But to the extent that the new naturalism is true, its pursuit seems certain to generate two great spiritual dilemmas.

The first is that no species, ours included, possesses a purpose beyond the imperatives created by its genetic history. The reflective person knows that his life is in some incomprehensible manner guided through a biological ontogeny, a more or less fixed order of life stages. He senses that with all the drive, wit, love, pride, anger, hope, and anxiety that characterize the species he will in the end be sure only of helping to perpetuate the same cycle. Poets have defined this truth as tragedy. Yeats called it the coming of wisdom:

Though leaves are many, the root is one;
Through all the lying days of my youth
I swayed my leaves and flowers in the sun;
Now I may wither into the truth.

The first dilemma, in a word, is that we have no particular place to go. The species lacks any goal external to its own biological nature. It could be that in the next hundred years humankind will thread the needles of technology and politics, solve the energy and materials crises, avert nuclear war, and control reproduction. The world can at least hope for a stable ecosystem and a well-nourished population. But what then? Educated people everywhere like to believe that beyond material needs lie fulfillment and the realization of individual potential. But what is fulfillment, and to what ends may potential be realized?

Let me state in briefest terms the basis of the second dilemma: innate censors and motivators exist in the brain that deeply and unconsciously affect our ethical premises; from these roots, morality evolved as instinct. If that perception is correct, science may soon be in a position to investigate the very origin and meaning of human values, from which all ethical pronouncements and much of political practice flow.

Which of the censors and motivators should be obeyed and which ones might better be curtailed or sublimated? These guides are the very core of our humanity. They and not the belief in spiritual apartness distinguish us from electronic computers. At some time in the future we will have to decide how human we wish to remain—in this ultimate, biological sense—because we must consciously choose among the alternative emotional guides we have inherited. To chart our destiny means that we must shift from automatic control based on biological processes to precise steering based on biological knowledge. Above all, for our own physical well-being if nothing else, ethical philosophy must not be left in the hands of the merely wise. Although human progress can be achieved by intuition and force of will, only hard-won empirical knowledge of our biological nature will allow us to make optimum choices among the competing criteria of progress.

The Great Divide

Robert Wright

There may be people who are devoted fans of both James Lovelock's Gaia hypothesis and Richard Dawkins' *The Selfish Gene*, but I suspect there aren't very many. For the worldviews embodied in these two men occupy opposite ends of a basic—perhaps *the* basic—spectrum of human thought. Some people call it the holist/reductionist spectrum, others the right-brain/left-brain spectrum. I propose that for present purposes we call it the weird/nonweird spectrum. A good number of Gaia proponents—the right-brain, holistic types (to oversimplify a bit)—seem to think that something weird is going on in this universe; most selfish-genists—the left-brain, reductionist types—don't.

This distinction isn't absolute, but it does have a rough statistical validity. Four out of five reductionists, I submit, are people who can find philosophical contentment in an utterly mechanistic universe, a universe devoid of souls, spirits, and other ethereal things—a universe, moreover, that may well be devoid of ultimate purpose. Four out of five holists, in contrast, are people who would prefer to believe that there is more to this universe than meets the eye—specifically, something suggestive of overarching design.

This great intellectual divide is perhaps less vast, and certainly less conspicuous, than it once was. During the early decades of this century, opponents of reductionism typically inhabited regions quite remote from the main currents of scientific discourse. In positing "élan vital" as the driving force behind evolution, Henri Bergson was not only invoking supernatural-sounding forces (a "vital impulse," a "current of consciousness"); he was also flatly rejecting the Darwinian view of evolution. Compare Bergson with today's holists. Though their idea of "self-organization" hasn't, as Lynn Margulis points out, entered the mainstream of biological thought, it is intended (a) to supplement, not displace, Darwinism and (b) as a basically material concept; when Ilya Prigogine speaks of "dissipative structures" and Humberto Maturana and Francisco Varela speak of "autopoeisis," they may incur the disdain of some physicists, chemists, and biologists, but they do mean to be describing physical principles.

Still, a universe with self-organization, like a universe with élan vital, is a universe that seems almost designed to create life; today's holists, like yesterday's, see the universe as a weirder, perhaps more purposeful, place than do the reductionists. Indeed, some reductionists might say that, however sober the new holists try to sound, they're just pouring old wine into new grape-juice bottles. And it's certainly true that holists do seem sometimes to be straining to conform fully with the technical specifications of modern science, lest they be dismissed as wooly-minded New Agers. Thus Lovelock, aware that the word "purpose" raises all sorts of hackles in scientific circles, defines Gaia as "an automatic, but not purposeful, goal-seeking system." (Meanwhile, of course, reductionists have also been known to make rhetorical overtures across the weird/nonweird chasm, waxing rhapsodic about the spiritual and aesthetic sides of scientific materialism lest they be dismissed as Godless bores. Card-carrying reductionist E. O. Wilson once candidly told an interviewer, in reference to his elegantly reflective book *On Human Nature*, "To the best of my ability, I made allusions to literature. I've used quotes from Yeats and Joyce and so forth.")

It is important, before proceeding further, to understand why Lovelock's avoidance of the word "purpose" isn't, in the final analysis, of any more substantive significance than Wilson's allusions to Yeats, and therefore won't insulate him against indictment for the high scientific crime of teleology. Suppose we grant him that the Gaia he describes really isn't "purposeful," but is merely, as he puts it, "goal-seeking." This distinction has an overly subtle sound to it, but presumably what Lovelock means is one of two things: (a) that "purposeful" systems *consciously* pursue goals, whereas "goal-seeking" systems may be unconscious (in which case "goal-*seeking*" is a questionable choice of terms), or (b) that "purposeful" systems are the product of design, whereas "goal-seeking" systems aren't. Goal-seeking systems, in other words, act *as if* they were consciously in pursuit of some goal, or *as if* they were designed for that purpose, though neither of these is necessarily true.

If you're tempted to believe that Lovelock's semantic microsurgery turns Gaia into a goddess that even a reductionist could love, stop and consider: What other things in this universe are "goal-seeking" (in the conventionally understood sense of utilizing feedback information to approach or maintain a particular state)? So far as I know, only two kinds of things: organic life (bacteria, humans, etc.), and things created by organic life (thermostats, robots, etc.). Now, the problem for Lovelock, as has been pointed out, is that we understand perfectly well how things like bacteria and humans and thermostats and robots came to seek goals: the bacteria and humans were created by a process, natural selection, that unconsciously endows systems with this tendency; the thermostats and robots were created by beings (us) that consciously endow systems with it. But what created Gaia? Reductionists, who so far have been able to explain in principle how every goal-seeking system in the universe came to be, are naturally going to be skeptical when confronted with a goal-seeking system whose creation, whose reason for being, is a mystery.

It's nice of Wilson and Lovelock to try so hard to speak each other's language, but let's face it: all their implicit protestations notwithstanding, their differing feelings about weirdness are indeed at the core of what divides them. E. O. Wilson can write beautifully and honestly about the quasi-religious rush he gets from science, but in the end he is a person who doesn't need religion in the traditional sense of the word; he doesn't need to believe in divine forces to find happiness. I suspect that Lovelock, in contrast, would rather like to. He is not that far from admitting as much when he writes, "Thinking of the Earth as alive makes it seem, on happy days, in the right places, as if the whole planet were celebrating a sacred ceremony."

Is it possible to reach a comparable state of bliss without Gaia? Can a strictly reductionist view of life on earth yield the sort of spiritual premiums Gaia gives? Maybe so. Here, for example, is one attempt to find happiness via the left brain: on Darwinian grounds alone, we are right to feel a fundamental kinship with all other living things; we are all products of a common creator, a creator of awesome ingenuity, awesome power (and, yes, awesome cruelty); the contours of all our lives—including our very physical configurations—have been shaped by various arbitrary factors but nonetheless possess a functional elegance still unrivalled in the world of artifacts. And our kinship with other *sentient* beings is even more profound: we have all been blessed and cursed, granted the capacities for pleasure and pain, sadness and joy. Look into a dog's eyes sometime. If you don't feel bound by a common predicament that is at once sublime and absurd, you're not looking close enough. And once you've mastered this, go out in the woods and bond with some deer and frogs.

There. How's that? Probably not good enough for most Gaia devotees. They won't settle for a sense of organic oneness couched in plain old Darwinian terms. They want to think that we and other forms of life are in not just the same predicament but the same organism. They want *real* bonding—with cement. But I don't think it's the bonding per se that holds the main attraction for them. I don't think it's the high degree of interspecies (that is, intrasuperorganic) kinship implied by Gaia that gives them their spiritual thrill; at bottom, I think, it's the mystery of how this giant global organism came to be. It is *because* the purpose (oops, I mean "goal-seeking tendency") purportedly exhibited by Gaia is uniquely inexplicable that they rejoice in it. For this suggestion of supernatural spookiness leaves open the possibility that the ruling power on this planet—or in this universe—is less arbitrary than natural selection; it leaves open the possibility that there's some *point* to this whole exercise in which all the deer and frogs and people are enmeshed.

Well, okay. Fair enough. You want overarching universal purpose? We'll see what we can do. Let's return to a conventionally reductionist Darwinian view of life and see if that won't provide some suggestive mysteries of its own.

Mystery no. 1: The miracle of sentience

There is no clear reason why it should feel like anything to be alive. Biologists have paid remarkably little attention to this fact—John Maynard Smith, one of the clearest thinkers in the discipline, is the only one I know of who has noted it—and yet it is a question that arises inescapably from a fundamentally reductionist view of evolution. If you believe that we are survival machines, robots programmed by our genes through an incredibly intricate flow of physical information (mediated largely by the brain, of course, and including massive informational input from the environment) and if you believe that this flow of information accounts for all of our behavior, then what is the need for pleasure and pain and sentience in general? It isn't good enough to say, "Oh, pain and pleasure are great motivators; they get you to eat food and avoid burning your hands." For if you truly are a conventional reductionist, you believe that these goals are accomplished through a sheerly physical flow of information—neuronal, hormonal, etc.—and that the accompanying sensations of pleasure and pain are mere epiphenomena, possessing roughly the relationship to the real dynamics of motivation that a car's shadow bears to the car.

So what's the answer? Why *does* it feel like something to be alive? Do the metaphysical laws of this universe dictate that complex information-processing be accompanied by sentience? Or that *organic* information-processing entail sentience? Or is it possible that the flow of physical organic information *isn't* exhaustively determinative? That, indeed, there *is* some immaterial metaphysical entity which guides behavior and is called free will? I have no idea. All I know is that we seem to have a bona fide mystery on our hands here. And what's more, a momentous mystery. For sentience, the experience of pleasure and pain, is what gives life meaning and gives moral questions their terrible weight. It is the reason that everything matters. Isn't it amazing that when natural selection created life, the metaphysical laws of this universe happened to toss in meaning as a freebie? If that isn't weird enough to leaven your existential despair, then I'm afraid you may be beyond help.

Mystery no. 2: The escalation of sentience

No respectable thinker in biology any longer believes that there is an inherent evolutionary impetus toward higher levels of physical organization and of sentience; there is no élan vital, and there is no sense in which human beings were in the cards from the beginning. On the other hand, it seems clear that the basic principles of organic economy (economies of scale, division of cellular labor, the advantages of behavioral flexibility, etc.), as mediated by natural selection, have often tended to drive life toward higher levels of organization and toward the more complex processing of information (notably in the form of bigger brains). This fact—mixed in with the fact that informational complexity involves high levels of sentience—means that evolution has often tended to create higher and higher degrees of consciousness, culminating (so far) in us, the *self*-conscious species. This has led to all sorts of profoundly meaningful things: the conscious understanding of the process that created us, the ability to wonder at its ultimate origin, the ability to consciously construct a moral code designed to elevate the quality of sentience on this planet, and, of course, the ability to feel pain and pleasure, sorrow and gladness, more complexly and subtly than any other species. I am tempted to say that natural selection was designed to generate meaningful life. But the word "designed" would get me in as much trouble as the word "purpose" gets Lovelock in—and rightly so. I'll leave the choice of words to you.

Mystery no. 3: How a process as ingenious as natural selection got started

Forget sentience; forget the fact that it has escalated through evolutionary time. The mere nuts-and-bolts physical facts of natural selection are abundantly amazing. That such breathtaking complexity could be created by so few basic principles is at least as awe-inspiring as the presence of a deity. Indeed, speaking of deities: on these grounds alone it is hard to believe that the whole process wasn't designed (whoops) by some intelligent being. But of course, the trouble with hypothesizing a creator of natural selection is that it doesn't simplify anything. After all, what created the creator?

We live in a world of profound and sweeping mystery, a world endowed with intelligence and with a kind of meaning and containing scattered clues, though no definitive evidence, of higher purpose. This much can be said without the help of Gaia, and not much more along these lines can be said with her help. So if she is going to earn a lasting place in Western discourse, it should be strictly on scientific merit; we shouldn't buy into the Gaia hypothesis for the sake of philosophical solace. This may (and, of course, may not) mean that we'll eventually lose her. But it also means that in philosophical terms, the loss won't be great.

Bibliography and Suggestions for Further Reading

I Is Earth Itself Alive?

This part drew upon selections from two books by James Lovelock, *Gaia: A New Look at Life on Earth* (Oxford University Press, 1979) and *The Ages of Gaia* (Norton, 1988). If you enjoyed Lovelock's writings as a most pleasurable way to learn science, I strongly encourage you to read these books and perhaps his two other popular works: *The Greening of Mars* (a science-fiction piece, published by St. Martin's Press, 1984) and *The Great Extinction* (Doubleday, 1983). The latter two were cowritten with Michael Allaby. If you enjoyed Lovelock's philosophical ramblings as well, *The Ages of Gaia* holds many more profound passages, which I simply had no room here to include.

For going deeper into the science and the personalities involved in the Gaia hypothesis, you might pick up a copy of Lawrence E. Joseph, *Gaia: The Growth of an Idea* (St. Martin's Press, 1990). Joseph finds a way to make even the straight science sections a very fine read; his biographical chapters are exquisite. If you are inclined to tackle scientific papers pertaining to the Gaia hypothesis, a good place to start is "Hands up for the Gaia Hypothesis,"a three-page invited commentary by Lovelock that appeared in the 8 March 1990 issue of the journal *Nature*. Next I recommend you peruse my own paper, which was published in *BioSystems* 23 (1990): 371–384. Called "Open Systems Living in a Closed Biosphere: A New Paradox for the Gaia Debate," the paper provides background on the genesis and subsequent uses of the Gaia hypothesis, with an extensive bibliography that will put you in touch with the key papers. (I got the idea for this paper while doing the background research for this anthology and found a scientist who was willing to help me develop the ideas into a form suitable for academic publication.) Another entry into the technical aspects of Gaia is through published proceedings of conferences. The MIT Press is now in the process of publishing several such volumes, including *The Science of Gaia,* edited by Stephen H. Schneider and Penelope J. Boston.

Respected critiques of Gaia include the book review "Is Nature Really Motherly?" by W. Ford Doolittle in the spring 1981 issue of *Coevolution Quarterly* (now published under the name *Whole Earth Review*). Richard Dawkins includes a short critique in his *Extended Phenotype* (W. H. Freeman, 1982). James W. Kirchner published powerful criticism of the Gaia hypothesis asserting that Gaia cannot be science because the hypothesis has not been stated in falsifiable terms. Kirchner's paper "The Gaia hypothesis: can it be tested?" was published first in *Reviews of Geophysics* 27 (1989): 223–225 and will soon be published in a conference volume on Gaia by The MIT Press, *The Science of Gaia.* It is a good place to explore the philosophy-of-science school launched by Karl Popper.

If you are interested in reading more about a highly publicized environmental problem of truly Gaian scale, the greenhouse effect, track down Stephen H. Schneider, *Global Warming* (Sierra Club Books, 1989). Following Schneider's work, a burst of books on global warming hit the bookstores; they are too numerous to list here.

If the Gaia hypothesis inspires you to think that human beings with our high-tech communications may become a kind of nervous system for Gaia, *The Global Brain* by Peter Russel (Houghton Mifflin, 1983) presents a blend of science and New Age spirituality. The classic work on the idea of an evolving global consciousness is, of course, *The Phenomenon of Man* by Teilhard de Chardin, published after his death and translated from the French (Harper and Row, 1959). Teilhard's worldview, however, is more anthropocentric than is Lovelock's view of Gaia (Lovelock views microbes as far more significant parts of the Gaian system). I personally find Teilhard's prose difficult, but the book has inspired a cult following. For a similar scientific and spiritual thesis, I prefer several of Julian Huxley's books: *Evolution in Action* (Harper, 1953), *Religion without Revelation* (Harper, 1957).

Eric Chaisson's *The Life Era: Cosmic Selection and Conscious Evolution* (Norton, 1987) is a more up-to-date version of some of the same ideas. It is very well written and offers an extended view of evolution that begins with the physical processes that formed the galaxies long before biological processes entered the scene. Likewise, Elisabet Sahtouris does a fine job of weaving a worldview (with a feminist perspective) out of the Gaia concept; James Lovelock wrote the foreword to her book, *Gaia: The Human Journey from Chaos to Cosmos* (Pocket Books, 1989). If Sahtouris's forays into issues of gender and science intrigue you, refer to *Reflections on Gender and Science* by Evelyn Fox Keller (Yale University Press, 1985).

An article by Frederick Turner that appeared in the August 1989 issue of *Harper's Magazine* draws upon Lovelock's work. In "Life on Mars: Cultivating a Planet and Ourselves," Turner sets out a bold proposal for unifying humankind's goals through an international effort to colonize Mars. For a twist on the theme of evolution, see Dorion Sagan, *Biospheres: Metamorphosis of Planet Earth* (McGraw-Hill/Bantam, 1990). Sagan pictures closed ecosystems as Gaia's technological offspring; they represent reproduction at a level higher than that of animality. Humans are not the brains of Gaia but are involved in the biosphere's reproduction, its generation. In this way, Sagan believes, individuality crops up at ever higher levels in evolution, and (at least at this moment) humans do have an important role to play. But the role is not a kind of Teilhardian noospheric world consciousness and transcendence but the physical business of propagating life beyond our home planet. Once that gets underway, we humans are dispensible.

If you enjoyed the section "The World's Biggest Membrane" (chapter 3), you may want to peruse the book from which it was drawn. *The Lives of a Cell* by Lewis Thomas (Viking/Bantam, 1974) won the National Book Award. He has several other collections of essays. All of Thomas's books are extremely popular, and you can find them in just about any library or bookstore.

"Songs for Gaia" appeared in Gary Snyder's *Axe Handles* (North Point Press, 1983). Snyder won a Pulitzer Prize for another of his books of poetry, *Turtle Island* (New Directions, 1974).

Full references for the other books and essays excerpted are provided within the text of part 1.

II Merged Beings

The profile of Lynn Margulis by Jeanne McDermott was drawn from "A biologist whose heresy redraws Earth's tree of life" in the August 1989 issue of *Smithsonian* magazine. Another excellent short biography, "One Woman and Her Theory," was written by Evelyn Fox Keller (who wrote the acclaimed biography of Barbara McClintock, *A Feeling for the Organism*); Keller's profile of Margulis appeared in the 3 July 1986 issue of *New Scientist.*

Lynn Margulis has written several technical books on the role of symbiosis in evolution and on the evolutionary developments in ancient bacteria that set the stage for plant and animal evolution. Most relevant here is her *Symbiosis in Cell Evolution* (W. H. Freeman, 1981). But the book I have excerpted here, *Microcosmos* (written with Dorion Sagan), is an extraordinary amalgam of all her major ideas geared for a general audience. *Microcosmos* (Summit Books, 1986) is probably the single best book I have read that combines the scientific story of the evolution of the cosmos with that of the evolution of life. It is called *Microcosmos* because the key events in biological evolution happened long before trilobites and brachiopods populated the Cambrian seas. The microscopic world of bacteria is the focus, and the book reveals the vital role played by these ubiquitous little beasts in keeping the Earth habitable for all forms of life.

The notion that symbiosis is a powerful force in evolution has generated a number of more philosophical books. *The Unfinished Universe* by Louise Young (Simon and Schuster, 1986) blends scientific facts with a philosophical worldview that resembles that of Teilhard de Chardin. Similarly, George Greenstein has recently written *The Symbiotic Universe* (Morrow, 1988). Meanwhile, very technical books and papers are being published on the role of symbiosis in evolution. The MIT Press will publish in 1991 *Symbiosis as a Source of Evolutionary Innovation: Speciation and Morphogenesis*, edited by Lynn Margulis and René Fester, the proceedings of a Bellagio conference in Italy.

The reason symbiosis gets such play today is that evolution theory did not always embrace the concept. Charles Darwin and especially subsequent theorists pointed to competition as the sole process by which some variations of life-forms survived and reproduced while others failed. Lynn Margulis has pointed out that cooperative interactions can provide symbiotic partners with an edge for survival. The first author to counter the competitive interpretation of evolution was the Russian scientist and anarchist Peter Kropotkin. His 1902 book *Mutual Aid* is a classic and is now published by Freedom Press: London, 1987. It is, moreover, a delight to read and provides extraordinary insight into the scientific and political arguments of the late nineteenth century. Ashley Montagu has chronicled the history of this debate in *Darwin: Competition and Cooperation* (Greenwood, 1952), and he dedicates this book to the memory of Kropotkin.

Chapter 6 is drawn from a classic book on the subject of individuality by Julian Huxley. Sadly, *The Individual in the Animal Kingdom* (Cambridge University Press, 1912) is currently out of print and very difficult to obtain even from libraries. Short excerpts are also drawn from two very recent books: *Population Biology and Evolution of Clonal Organisms*, edited by Jeremy B. C. Jackson, Leo W. Buss, and Robert E. Cook (Yale University Press, 1984), and *The Evolution of Individuality* by Leo W. Buss (Princeton University Press, 1987, pages 25, 174, 175, 194, 195). Both are highly technical. Nevertheless, the latter is the most mind-jarring work I have seen on the evolution of individuality, and I do recommend it.

Daniel Janzen presents a lively introduction to the problem of individuality in his "What Are Dandelions and Aphids?" which appeared in *American Naturalist* (111 (1977): 586–589. David Hull and Richard Dawkins also explore issues of evolutionary biology that pertain to individuality in the anthology *Genes, Organisms, Populations: Controversies over the Units of Selection*, edited by Robert N.

Brandon and Richard M. Burian (MIT Press, 1984). Although geared for a technical audience, the essays by Hull ("Units of Evolution") and Dawkins ("Replicators and Vehicles") are profoundly interesting and not too esoteric.

III A Systems View of Life

Chapter 7 draws on two books by Arthur Koestler: *The Ghost in the Machine* (Macmillan, 1967) and *Janus: A Summing Up* (Random House, 1978). Probably the best book to peruse is the latter, as Koestler produced it to bring together all the major ideas of his lengthy career. Among other things, it is a kind of manifesto for the holistic approach to scientific and other research. Koestler is editor of *Beyond Reductionism* (Macmillan, 1969), a compilation of presentations made at a 1968 conference by prominent scientists (including Paul Weiss and Ludwig von Bertalanffy) who utilize holistic or nonreductionistic approaches in their work. Koestler's quotation of Ruth Sager is drawn from "Genes Outside the Chromosomes" in the January 1965 issue of *Scientific American*. His quotation of Lewis Thomas is drawn from "An Earnest Proposal," which appears in Thomas's *The Lives of a Cell* (copyright 1973 by The Massachusetts Medical Society, reprinted by permission of Viking Penguin, a division of Penguin Books USA Inc.). Koestler's quotation of Howard H. Pattee is drawn from *Towards a Theoretical Biology*, edited by C. H. Waddington (Edinburgh, 1970). His quotation of Joseph Needham is drawn from *Order and Life* by Needham (New Haven, Conn., 1936).

A number of scientists have produced technical works on the use of hierarchy theory in scientific research. A good place to begin is *Hierarchy: Perspectives for Ecological Complexity* by Tim Allen and T. B. Starr (University of Chicago Press, 1982). I found this book very helpful while working on my paper on Gaia. It helped clarify my ideas pertaining to ecological systems, including the biosphere, and gave me concrete tools for determining system boundaries. Two early essays on the subject are "New Concepts in the Evolution of Complexity: Stratified Stability and Unbounded Plans," by Jacob Bronowski (which appeared in a 1970 issue of the journal *Zygon*), and "The Architecture of Complexity," by Herbert A. Simon, *Proceedings of the American Philosophical Society*, 106 (1962): 467–482. (You may recall that Koestler attributes his tale of the two watchmakers to Simon.)

Chapter 8 is drawn from Ludwig von Bertalanffy's classic *Problems of Life* (Wiley, 1952). I rank this book among a mere dozen books (including Lovelock's *Gaia* and Dawkins' *Selfish Gene*) that have launched me into the highest orbits of intellectual heaven for a few days. If you are intrigued with either the metaphysics or the methodology of general systems theory, there are a number of books I strongly recommend. Gerald M. Weinberg produced *An Introduction to General Systems Thinking* (Wiley, 1975). Geared for practical use, it helped me understand how I think and the kinds of approaches to problem solving that have served me best in my various careers. Ludwig von Bertalanffy penned *General System Theory* (1968), but I prefer his *Problems of Life*. Ervin Laszlo edited *The Relevance of General Systems Theory*, (George Braziller, 1972). Contributors include Anatol Rapoport and Kenneth Boulding. Together with von Bertalanffy, these two men rank as key early contributors to systems theory. Robert Wright presents a powerful biographical essay on Boulding and the spiritual ramifications of systems theory in *Three Scientists and Their Gods* (Times Books, 1988).

Joseph Needham wrote a beautiful book that anticipates the ideas of the systems theorists; *Order and Life* (New Haven, Conn., 1936) is a classic holistic work. Finally, Stephen Jay Gould does a splendid job of putting the holist/reductionist debate in perspective in his January 1984 column in *Natural History*.

IV *Game Theory and the Evolution of Cooperation*

This part is drawn almost entirely from Robert Axelrod's *The Evolution of Cooperation* (Basic Books, 1984). (I also include a chunk of an essay by Douglas Hofstadter, which appeared in his *Metamagical Themas*, also published by Basic Books, 1985). But there is a great deal more to Axelrod's book than I was able to use in this anthology. I have never been interested in war stories, but I was entranced while reading Axelrod's chapter presenting a game-theoretic analysis of why troops fighting in the trenches during World War I developed such uncombat like behavior as "live and let live." In another chapter Axelrod draws upon his knowledge of game theory to suggest ways for politicians and political activists to bring about a more peaceful and conciliatory world. Hofstadter's essay too is far richer than the short excerpt presented here. He ruminates on game-theoretic implications of everyday affairs like getting your car fixed, driving it in heavy traffic, and boarding a jet. He also ranges into a whole other scale of concern, with thoughts on prisoner behavior in Nazi concentration camps and military arms races.

The evolution theorist John Maynard Smith has written several books and essays for a mixed professional and general audience that portray game theory as a tool for biological understanding. One book is titled *Evolution and the Theory of Games* (Cambridge University Press, 1982). Richard Dawkins, in his *The Selfish Gene* (especially the 1989 edition), also does an excellent job presenting the practical uses to which biologists put game theory. If you are curious about the classic formulation of the Prisoner's Dilemma, see Anatol Rapoport's *Prisoner's Dilemma* (University of Michigan Press, 1965).

Axelrod refers to a paper by Robert L. Trivers: "The Evolution of Reciprocal Altruism," *Quarterly Review of Biology* 46 (1971): 35–57. He also recommends Trivers, *Social Evolution* (Menlo Park Calif.: Benjamin/Cummings, 1985). Axelrod's own latest technical writings are "The Further Evolution of Cooperation," coauthored with Douglas Dion, which appeared in *Science* 242 (1988): 1385–1390, and a chapter in Lawrence Davis, ed., *Genetic Algorithms and Simulated Annealing* (Los Altos, Calif.: Morgan Kaufman, 1987).

V *Nature, Nurture, and Sociobiology*

The part begins with a profile of the founder of sociobiology, Edward O. Wilson, drawn from *Three Scientists and Their Gods*, by Robert Wright (Times Books, 1988). This book captivated me like a good novel, and I highly recommend that readers sample the original. (Robert Wright covers the polymaths Kenneth Boulding and Ed Fredkin as well as Wilson.)

Wilson himself has written or cowritten half a dozen books on sociobiology, but most are technical. Wilson's essay in the *New York Times Magazine* (12 October 1975), which appears in this anthology, is really the best nontechnical presentation I have found. You may, however, want to take a look at the monumental treatise that started the whole field, *Sociobiology: The New Synthesis* (Harvard University Press, 1975). The book that won Wilson a Pulitzer Prize, and which generated a torrent of criticism, is *On Human Nature* (Harvard University Press, 1978). It is fascinating reading (excerpts from this book appear in chapter 17), and Wilson ranges widely—he even poses scientific materialism as an alternative to established religions. It is also a book that is sure to incite your passions—for or against.

If you are intrigued by Wilson the scientist (he has made important theoretical contributions in a range of biological fields), I recommend that you read his semiautobiographical *Biophilia* (Harvard University Press, 1984). I particularly enjoyed his story about how he and Robert MacArthur developed the theory of island biogeography, which leads Wilson to ruminate on the similarities and differences between artists and scientists.

A lot of other books have been written by proponents of sociobiology. Next to Wilson, one of the most respected theorists is Richard D. Alexander. His *The Biology of Moral Systems* (De Gruyter, 1987) is accessible to the general reader. David Barash also writes popular-science books from a sociobiological perspective. His most recent book is *The Tortoise and the Hare: Culture, Biology, and Human Nature* (Viking, 1986).

If you want to investigate the classic technical papers on which the field is based, see the references on game theory and reciprocal altruism mentioned for part IV. In addition, kin selection was first expounded by William Hamilton in a two-part paper, "The Genetical Evolution of Social Behavior," that appeared in the *Journal of Theoretical Biology* 7 (1964): 1–52. John Maynard Smith published "The Evolution of Behavior" in *Scientific American* 239 (1978): 176–191; it explores all these ideas and contributors in nontechnical terms.

A lot of books have been written by opponents of sociobiology as well. I have drawn from *Not in Our Genes* (Pantheon, 1984) by Richard C. Lewontin, Steven Rose, and Leon J. Kamin. In addition to being well written, the work delves into history and philosophy, and for this reason it is an excellent introduction to one particular worldview (dialectical materialism) that rejects the kind of reductionism inherent in sociobiology.

If you are intrigued by this worldview, you may wish to sample Lewontin's *The Dialectical Biologist* (cowritten with Richard Levins, Harvard University Press, 1985). Or you may want to go right to the source: Frederick Engels, *Dialectics of Nature*, which is one of the classic works that Marxist scientists and politicians turn to for guidance. Dialectical materialism is an immense subject with worldscale effects. I have found Loren Graham's *Science, Philosophy, and Human Behavior in the Soviet Union* (Columbia University Press 2nd ed., 1987) to be helpful in providing a fair and fascinating view of the historical context as well as the content.

For an additional critique of biological reductionism and a dialectical alternative, see Steven Rose's *Molecules and Minds* (Milton Keynes; England: Open University Press, 1987), or a book that he edited with Hilary Rose, *The Radicalisation of Science: Ideology of/in the Natural Sciences* (Macmillan, 1976). Leon J. Kamin's classic work is *The Science and Politics of IQ* (Potomac, Md.: Erlbaum, 1974).

Excerpts from two other book-length critiques of sociobiology are presented in chapter 13. *Vaulting Ambition* by Philip Kitcher (MIT Press, 1987) analyzes sociobiology from a non-Marxist perspective. Much of it is a bit too technical for my purposes here, but it is nevertheless excellent. Stephen Jay Gould judges it "the best dissection ever published on the logic and illogic (mostly the latter) of sociobiology." The passage by Ashley Montagu was drawn from *Sociobiology Examined*, which he edited (Oxford University Press, 1980).

A provocative debate occurred in the pages of *Bioscience* 26 (March 1976):182–183. A group of more than thirty scientists and other professionals (including Stephen Jay Gould and Richard Lewontin) published a joint piece that attacked the very foundations of sociobiology theory. Both the critique and Edward O. Wilson's response (published in the same issue) are forthright, erudite, and persuasive.

VI Selfish Genes

Chapters 14 and 15 are Richard Dawkins through and through. If you want to learn more about the selfish gene theory, read Dawkins and more Dawkins. The excerpts were all drawn from his *The Selfish Gene* (Oxford University Press, 1976). I suggest that you acquire the 1989 edition as Dawkins has added two chapters to account for advances in game theory (notably Robert Axelrod's work) and to account for Dawkins' own work on a new theory called "the extended phenotype." Though far more technical than *The Selfish Gene*, I found *The Extended Phenotype* (W. H. Freeman, 1982) to be even more mind-jarring. Dawkins' *The Blind Watchmaker* (Norton, 1987) moves on to a different theme in evolution, and it is written for a popular audience. Dawkins was awarded the Royal Society of Literature prize and the *Los Angeles Times* literary prize for this book.

If you would like to read more about the idea of "cultural evolution," which Dawkins introduces, you might begin with *Mankind Evolving*, a classic work by the evolution theorist Theodosius Dobzhansky (Yale University Press, 1962).

If you would like to read more about the origins of life, I recommend that you begin with Freeman Dyson's *The Origins of Life* (Cambridge University Press, 1985). Dyson is an extraordinary writer, and because he is a physicist and mathematician by training, his foray into biology takes nothing for granted. He first presents a history of ideas on the origins of life and then makes some startling suggestions of his own. Dyson and Dawkins both specifically mention the work of A. G. Cairns-Smith, who hypothesized that clay minerals played a key role in the origin of life. Cairns-Smith wrote a popular book on his ideas: *Seven Clues to the Origin of Life* (Cambridge University Press, 1985). Sydney Fox's research has long challenged the "nucleic acid" focus of colleagues of Dawkins; Fox posits that amino acids (proteins) played the key role. For this viewpoint, consult Fox's *The Emergence of Life* (Basic Books, 1988). Robert Shapiro has produced a popular account that tears into everybody's theory: *Origins: A Skeptic's Guide to the Creation of Life on Earth* (Summit, 1986).

Two scientific papers on selfish or junk DNA are excerpted in chapter 16. Both appeared in the same issue of the prestigious British journal *Nature* 284 (1980): 601–607.

I have read several popular essays that dispute Dawkins' selfish gene theory, but the only ones that I have seen and would recommend are by Stephen Jay Gould; see his columns in the December 1977 and November 1981 issues of *Natural History*. Some other scientists and philosophers have infused their criticisms with so much venom that I am reluctant to recommend them. As I mention in the text, perhaps the best critique can be acquired by referring to the essays in chapters 7 and 8.

If you are willing to get into the heavy stuff, the respected philosopher of science David Hull has written a powerful piece in the *Annual Review of Ecological Systems* 11 (1980): 311–332. Here he suggests that "interactors" should replace Dawkins' terminology "survival machines." Hull's suggestion is more than semantic, however, and you will be drawn into the now-heated "units of selection" debate. Hull's *Science as a Process* (University of Chicago, 1988) also has a chapter on this topic, from which the excerpt was drawn. If this debate intrigues you, you may wish to peruse volumes of *Philosophy of Science* during the 1980s, as many articles on this question have appeared there. But perhaps the best introduction to this debate is *Genes, Organisms, Populations: Controversies over the Units of Selection*, edited by Robert N. Brandon and Richard M. Burian (MIT Press, 1984). In addition to essays by Dawkins and Hull, this anthology includes contributions from other prominent evolution theorists or philosophers like Ernst Mayr, George C. Williams, Elliott Sober, Richard Lewontin, William D. Hamilton, and John Maynard Smith.

Readings in the Philosophy of Science

Speaking of philosophy of science, if this anthology sets you to wondering about the nature of scientific inquiry and the interactions of science and culture, a whole raft of books awaits you. A highly readable classic in this field is Thomas Kuhn's *The Structure of Scientific Revolutions* (University of Chicago Press, 1962). For expansion and criticism of Kuhn's thesis, *Criticism and the Growth of Knowledge*, edited by Imre Lakatos (Cambridge University Press, 1970), introduces the ideas of more than half a dozen philosophers.

More recently Larry Laudan has published some intriguing new ideas that are stimulating debate. Consult his *Science and Values* (1984) and *Progress and Its Problems* (1977). Both came out of the University of California Press. A good overview can be had in *Scientific Revolutions*, edited by Ian Hacking (Oxford University Press, 1981). There you will be introduced not only to the ideas of Kuhn, Lakatos, and Laudan but also to the other most important philosophers of science: Karl Popper, Paul Feyerabend, Dudley Shapere, and Hilary Putnam. Other useful anthologies include *Scientific Knowledge*, edited by Janet A. Kourany (Belmont Calif.: Wadsworth Publishing Co., 1987), and *Introductory Readings in the Philosophy of Science*, edited by E. D. Klemke et al. (Buffalo, N.Y.: Prometheus Books, 1988)

If you are more interested in the sociology than the philosophy of science, you will do well to read Loren Graham's captivating book *Between Science and Values* (Columbia University Press, 1981). Francois Jacob produced a great book on the same theme: *The Possible and the Actual* (University of Washington Press, 1982). To round out the perspective, you may wish to read one or more of Stephen Toulmin's works, notably *The Return to Cosmology* (University of California Press, 1982) and *Foresight and Understanding* (Indiana University Press, 1961).

VII Reflections

The essay by Edward O. Wilson is drawn from his book *On Human Nature* (copyright 1978 by the President and Fellows of Harvard College, Harvard University Press). The essay by Lynn Margulis is adapted from "Big Trouble in Biology" in *Doing Science: The Reality Club 2*, edited by John Brockman, (Prentice-Hall). For further reading, Margulis recommends "Origins of Life: An Operational Definition," by G. R. Fleischaker, *Origins of Life and Evolution of the Biosphere* 20 (1990): 127–137.

Sources of Illustrations

Chapter 1

James Lovelock. Photograph copyright 1989 by Sandy Orchard. Courtesy of James Lovelock.

Wasp nest, artist unknown. Illustration printed in Jim Harter, *Animals: 1419 Copyright-Free Illustrations of Mammals, Birds, Fish, Insects, etc.* (Dover Publications, 1979).

The False Mirror by René Magritte, 1928. Oil on canvas, 54 x 80.9 cm. Collection of The Museum of Modern Art, New York. Purchase.

Chapter 2

"Radiolarians" by Ernst Haeckel, late nineteenth century. Illustration printed in Ernst Haeckel, *Art Forms in Nature* (Dover Publications, 1974).

Sky above Clouds IV by Georgia O'Keeffe, 1965. Oil on canvas, 243.8 x 731.5 cm. Gift of Georgia O'Keeffe; restricted gift of Paul and Gabriella Rosenbaum Foundation, 1983.821. Photograph copyright 1990 by The Art Institute of Chicago. All rights reserved.

Chapter 3

Earth rise. Photograph courtesy of the National Aeronautics and Space Administration.

The Dream by Henri Rousseau, 1910. Oil on canvas, 204.5 x 298.5 cm. Collection of The Museum of Modern Art, New York. Gift of Nelson A. Rockefeller.

"Interior of primeval forest on the Amazons" by Henry Walter Bates, ca. 1850. Illustration appears in Henry Walter Bates, *Naturalist on the River Amazon.* Copyright 1962 by The Regents of the University of California.

Chapter 4

Lynn Margulis working in the lab. Photograph copyright by Fred Le Blanc. Courtesy of Lynn Margulis.

Sperm tails in cross-section. Photograph copyright by Dr. Don Fawcett/Dr. David Phillips/Science Source/Photo Researchers.

Chapter 5

Krishna as Navagunjara. Miniature, Kulu ca. 1700, 150 x 220 mm. Private collection. Reprinted from *Krishna: The Divine Lover* (copyright 1982 by Edita S. A.) with permission by David R. Godine, Publisher Inc., Boston.

"Pan" by Margaret Evans Price in *A Child's Book of Myths and Enchantment Tales*. Copyright 1924, 1952 by Checkerboard Press, a division of Macmillan, Inc. All rights reserved. Used by permission.

Five Dreamings by Michael Nelson Tjakamarra, assisted by Marjorie Napaljarri, 1984. Papunya, Central Australia. Acrylic on canvas, 122 x 182 cm. Permission of Aboriginal Artists Agency, 12 McLaren Street, North Sydney 2060.

Chapter 6

Procession of Monks, a Japanese Zen painting by Nantembō, 1924. Ink on paper, each 51 x 11.75 in. From the collection of Dr. Kurt Gitter (New Orleans); courtesy of Dr. Gitter.

Sky and Water I by M. C. Escher, 1938. Woodcut, 44 x 44 cm. Copyright 1990 by M. C. Escher Heirs/Cordon Art, Baarn, Holland.

"Sea Anemones" by Ernst Haeckel, late nineteenth century. Illustration appears in Ernst Haeckel, *Art Forms in Nature* (Dover Publications, 1974).

Aspen clone. Photograph copyright 1989 by Connie Barlow.

Chapter 7

Pantheon, interior of portico, plate 4 from *Veduta di Roma,* Giovanni Battista Piranesi, mid eighteenth century. Collection of The Metropolitan Museum of Art, Rogers Fund, 1941 [41.71.1 (17)]. All rights reserved, The Metropolitan Museum of Art.

Root nodules by Marcello Malpighi, Omnia Opera, London 1686. Illustration reprinted from *The Discovery of Nature* by Albert Bettex (copyright 1965 by Droemersche Verlagsanstalt Th. Knaur Nachf., Munich, Germany, published by Simon & Schuster).

Chapter 8

A collage assembled by Connie Barlow, 1990, from *Metamorphosis II* and *Stars* by M. C. Escher (copyright 1990 by M. C. Escher Heirs/Cordon Art, Baarn, Holland); "Chameleons," an illustration in Jim Harter, *Animals: 1419 Copyright-Free Illustrations of Mammals, Birds, Fish, Insects, etc.* (Dover Publications, 1979).

Bamboo, by Hsü Wei, sixteenth century. Section of a handscroll, ink on paper, 535.5 x 32.5 cm. Courtesy of the Freer Gallery of Art, Smithsonian Institution, Washington, D.C.

In the Mountains by Albert Bierstadt, 1867. Oil on canvas, 36.2 x 50.2 in. Wadsworth Atheneum, Hartford. Gift of John J. Morgan in memory of his mother, Juliet Pierpont Morgan.

Chapter 9

Butchering at Gambell, stone lithograph by Rie Muñoz, 1974. 20.25 x 13 in. Courtesy of Rie Muñoz, Ltd. (Juneau, Alaska).

The Cardsharps by Michelangelo Merisi da Caravaggio, 1594. Oil on canvas, 94.7 x 132 cm. Courtesy of Kimbell Art Museum, Fort Worth, Texas.

Chapter 10

Always More by Jean-Michel Folon, 1983. Cut and pasted paper, watercolor, colored pencils, and pencil. Copyright 1983 by Folon. Courtesy of John Locke Studios, Inc., New York.

Chapter 11

E. O. Wilson at the Costa Rica rainforest site. Photograph copyright by Bert Hölldobler. Courtesy of E. O. Wilson.

A queen of the leafcutter ant *Atta cephalotes*. Photograph copyright by Professor Carl W. Rettenmeyer, University of Connecticut.

Chapter 12

A view of the Iwo Jima Memorial. Official U.S. Navy photograph.

Mosaic II by M. C. Escher, 1957. Lithograph, 31.5 x 37.2 cm. Copyright 1990 by M. C. Escher Heirs/Cordon Art, Baarn, Holland.

Prehistoric rock painting discovered in Arnehm Land, Australia, "Hooked Stick" period. Faded dark red, maximum length of figure at left: 32 cm. Rendered by Eugene Beckes after figure 71 in Darrell Lewis, *The Rock Paintings of Arnhem Land, Australia*. Copyright 1988 by Darrell Lewis.

Les Demoiselles d'Avignon by Pablo Picasso, 1907. Oil on canvas, 243.9 x 233.7 cm. Collection of The Museum of Modern Art, New York. Acquired through the Lillie P. Bliss Bequest.

Chapter 13

"The Elephant's Child" by Rudyard Kipling.

Terraced rice agriculture in Bali. Photograph copyright 1990 by Michael Rampino. Courtesy of Michael Rampino.

Chapter 14

The gramineous bicycle garnished with bells the dappled fire damps and the echinoderms bending the spine to look for caresses by Max Ernst, 1920/1921. Botanical chart altered with gouache, 74.3 x 99.7 cm. Collection of The Museum of Modern Art, New York. Purchase.

My Grandparents, My Parents, and I (Family Tree) by Frida Kahlo, 1936. Oil and tempera on metal panel, 30.7 x 34.5 cm. Collection of The Museum of Modern Art, New York. Gift of Allan Roos, M.D., and B. Mathieu Roos.

Three generations of Dawkinses. Photographs of C. G. E. Dawkins and C. J. Dawkins courtesy of Richard Dawkins. Photograph of Richard Dawkins copyright 1990 by Lisa Lloyd.

Chapter 15

"The heavy infantry of the rebellious insects" by Grandville, mid nineteenth century. Lithograph appears in *Fantastic Illustrations of Grandville*. Copyright 1974 by Dover Publications.

Chapter 16

Plant gall by Marcello Malpighi, Omnia Opera, London 1686. Illustration reprinted from *The Discovery of Nature*, by Albert Bettex (copyright 1965 by Droemersche Verlagsanstalt Th. Knaur Nachf., Munich, Germany, published by Simon & Schuster).

Barlow with pet salamander. Photograph copyright 1989 by Tyler Volk.

Chapter 17

Five Kingdoms Hand (designed by Dorion Sagan), cover design for *Five Kingdoms* by Lynn Margulis and Karlene V. Schwartz, copyright 1982 by W. H. Freeman & Company. Courtesy of Lynn Margulis.

Authors

Richard D. Alexander is professor of zoology and curator of insects at the University of Michigan and is a member of the National Academy of Science. His sociobiological writings include *The Biology of Moral Systems* and *Darwinism and Human Affairs.*

Robert Axelrod is Distinguished University Professor of Political Science and Public Policy at the University of Michigan. A MacArthur Fellow, he is known for his work in game theory. He is author of *The Evolution of Cooperation.*

Ludwig von Bertalanffy, deceased, was a biologist. He is known today as the father of general systems theory. His writings include *General System Theory*, *Problems of Life*, and *Robots, Men, and Minds.*

Leo W. Buss is a biologist at Yale University. A MacArthur Fellow, he is author of *The Evolution of Individuality* and is coeditor (with J. B. C. Jackson and R. E. Cook) of *Population Biology and Evolution of Clonal Organisms.*

Joseph Campbell, deceased, was an expert on mythology. Bill Moyers produced a television show and a book of conversations with Campbell: *The Power of Myth.*

Francis Crick, F.R.S., is Distinguished Research Professor at the Salk Institute for Biological Studies in San Diego. He was corecipient of the 1962 Nobel Prize in physiology for his discovery (with James Watson) of the helical structure of DNA. His writings include *Of Molecules and Men* and *Life Itself.*

Richard Dawkins is an ethologist and evolution theorist at Oxford University. Originator of the selfish gene theory, he is author of *The Selfish Gene*, *The Extended Phenotype*, and *The Blind Watchmaker* (which earned him a Royal Society of Literature Award and the Los Angeles Times Literary Prize).

W. Ford Doolittle is a molecular biologist in the Department of Biochemistry at Dalhousie University, Nova Scotia. His work includes research on selfish DNA.

Stephen Jay Gould is professor of geology at Harvard University. A MacArthur Fellow, he specializes in paleontology and evolutionary biology. He is also an acclaimed essayist, whose books include *Wonderful Life: The Burgess Shale and the Nature of History*, *The Flamingo's Smile*, and *Ontogeny and Phylogeny.*

William D. Hamilton, F.R.S., is a zoologist at Oxford University. He originated the theory of kin selection.

Douglas Hofstadter is professor of cognitive science and computer science at Indiana University. He is author of *Metamagical Themas* and *Gödel, Escher, and Bach*, the latter bringing him a Pulitzer Prize.

David Hull is professor of philosophy at Northwestern University. His books include *The Metaphysics of Evolution*, *Science as a Process*, and *Philosophy of Biological Science*.

Julian Huxley, deceased, was a British biologist and evolution theorist. In the first half of the twentieth century he helped forge a synthesis of Darwin's theory of natural selection and Mendel's insights on genetics. He was a prolific writer; his books include *Evolution: The Modern Synthesis*, *The Individual in the Animal Kingdom*, and *Religion without Revelation*.

Lawrence E. Joseph is a science writer and author of *Gaia: The Growth of an Idea*.

Leon J. Kamin is professor and chair of the Psychology Department at Northeastern University in Boston. His writings include *The Science and Politics of IQ*, and he is coauthor of *Not in Our Genes*.

Philip Kitcher is a philosopher of science at the University of California, San Diego. Specializing in biology, his books include *Vaulting Ambition: Sociobiology and the Quest for Human Nature* and *Abusing Science: The Case against Creationism*.

Arthur Koestler, deceased, was a writer. His science books include *Janus: A Summing Up*, *The Ghost in the Machine*, and *The Sleepwalkers*.

Richard C. Lewontin is an evolutionary geneticist at Harvard University. His books include *The Genetic Basis of Evolutionary Change*, *Human Diversity*, and *Not in Our Genes* (with Steven Rose and Leon J. Kamin).

James Lovelock, F.R.S., is an independent scientist and inventor who lives in Cornwall, England. Originator of the Gaia hypothesis, his books include *Ages of Gaia* and *Gaia: A New Look at Life on Earth*.

Francesca Lyman is a science writer and author of *The Greenhouse Trap*.

Jeanne McDermott is a science writer and author of *The Killing Minds: The Menace of Biological Warfare*.

Lynn Margulis is a cell biologist and expert in microbial evolution and ecology at the University of Massachusetts, Amherst, and is a member of the National Academy of Science. An expert on symbiosis, she is the author of *Symbiosis in Cell Evolution*, *Early Life*, *Five Kingdoms* (with K. V. Schwartz), and *Microcosmos* (with Dorion Sagan).

John Maynard Smith, F.R.S, is an evolution theorist at the University of Sussex. His books include *Did Darwin Get It Right?* and *Evolution and the Theory of Games* and *The Evolution of Sex*.

Richard Monastersky is a journalist specializing in earth sciences for the magazine *Science News*.

Ashley Montagu is an anthropologist and writer who has been associated with several universities in the United States during his lengthy career. In addition to textbooks on anthropology, his numerous publications include *Culture and the Evolution of Man*, *Darwin: Competition and Cooperation*, and *Sociobiology Examined.*

Leslie Orgel, F.R.S., is a senior fellow at the Salk Institute for Biological Studies in San Diego. He is author of *Origins of Life: Molecules and Natural Selection.*

Steven Rose is a neurobiologist and chair of the Department of Biology at The Open University in England. He is author of *The Conscious Brain*, of *Molecules and Minds*, and is coauthor of *Not in Our Genes.*

Dorion Sagan is a science writer. He is author of *Biospheres: Metamorphosis of Planet Earth* and is coauthor of several books with Lynn Margulis, including *Microcosmos.*

Carmen Sapienza is a geneticist and head of the Laboratory of Developmental Genetics at the Ludwig Institute for Cancer Research in Montreal.

Gary Snyder is a poet and writer living in California. His books of poetry include *Axe Handles* and *Turtle Island* (the latter won a Pulitzer Prize).

Lewis Thomas is emeritus president of the Memorial Sloan-Kettering Cancer Center, a professor of medicine and pathology at Cornell University, and a member of the National Academy of Science. He is also an acclaimed essayist. His first collection of essays, *The Lives of a Cell*, earned him a National Book Award.

Edward O. Wilson is a behavioral biologist and evolution theorist at Harvard University and is a member of the National Academy of Science. Winner of the National Medal of Science and the father of sociobiology, he has written *Sociobiology: The New Synthesis*, *The Insect Societies*, and *On Human Nature* (the last was awarded a Pulitzer Prize).

Robert Wright is author of *Three Scientists and Their Gods.* He is a writer and a senior editor of *The New Republic.*

Index